SCIENTIFIC AND TECHNICAL ASSESSMENTS
OF ENVIRONMENTAL POLLUTANTS

Kepone/Mirex/
Hexachlorocyclopentadiene:
An Environmental Assessment

A report prepared by the
Panel on Kepone/Mirex/Hexachlorocyclopentadiene
of the Coordinating Committee for Scientific and Technical
Assessments of Environmental Pollutants

Environmental Studies Board
Commission on Natural Resources
National Research Council

NATIONAL ACADEMY OF SCIENCES
Washington, D.C.   1978

NOTICE

The project that is the subject of this report was approved by the Governing Board of the National Research Council, whose members are drawn from the Councils of the National Academy of Sciences, the National Academy of Engineering, and the Institute of Medicine. The members of the Panel responsible for the report were chosen for their special competences and with regard for appropriate balance.

This report has been reviewed by a group other than the authors according to procedures approved by a Report Review Committee consisting of members of the National Academy of Sciences, the National Academy of Engineering, and the Institute of Medicine.

This study was supported by the Office of Health and Ecological Effects, Office of Research and Development, U.S. Environmental Protection Agency, Contract No. 68-01-3253.

Library of Congress Catalog Card Number 78-58517

International Standard Book Number 0-309-02766-7

Available from

Printing and Publishing Office
National Academy of Sciences
2101 Constitution Avenue, N.W.
Washington, D.C. 20418

Printed in the United States of America

PANEL ON KEPONE/MIREX/HEXACHLOROCYCLOPENTADIENE

Robert J. Livingston (Chairman), Florida State University
Fumio Matsumura, Michigan State University
Gary M. Williams, American Health Foundation
Ian C.T. Nisbet, liaison with Coordinating Committee,
    Massachusetts Audubon Society

Consultant:

Charles C. Brown, National Cancer Institute

Staff Officer:

Lawrence C. Wallace, Environmental Studies Board

COORDINATING COMMITTEE FOR
SCIENTIFIC AND TECHNICAL ASSESSMENTS
OF ENVIRONMENTAL POLLUTANTS

Ian C.T. Nisbet (Chairman), Massachusetts Audubon Society
Ralph C. d'Arge, University of Wyoming
John W. Berg, The Colorado Regional Cancer Center
Russell F. Christman, University of North Carolina
Cyril L. Comar, Electric Power Research Institute
Eville Gorham, University of Minnesota
Robert C. Harriss, Florida State University
Delbert D. Hemphill, University of Missouri
Margaret Hitchcock, Yale University
Robert J. Moolenaar, Dow Chemical Company
Jean M. Morris, E.I. duPont de Nemours & Company
Peter N. Magee, Fels Research Institute

iii

Ex Officio Members (Panel Chairmen):

Julian B. Andelman, Panel on Low Molecular Weight
    Halogenated Hydrocarbons, University of Pittsburgh
Patrick L. Brezonik, Panel on Nitrates, University of
    Florida
Frank M. D'Itri, Panel on Mercury, Michigan State University
Robert J. Livingston, Panel on
    Kepone/Mirex/Hexachlorocyclopentadiene, Florida State
    University

Staff:

Edward Groth III, Environmental Studies Board
Adele L. King, Environmental Studies Board
Charles R. Malone, Environmental Studies Board,
(June 1975-December 1976)
Lawrence C. Wallace, Environmental Studies Board

ENVIRONMENTAL STUDIES BOARD

CONTENTS

PREFACE

In early 1975 the U.S. Environmental Protection Agency asked the National Research Council (NRC) of the National Academy of Sciences to prepare a series of comprehensive scientific and technical assessment documents on selected multimedia environmental pollutants. These documents were to be used by EPA as a basis for preparing Scientific and Technical Assessment Reports (STARs), which in turn would be used as the scientific and technical basis for possible regulatory action on these pollutants.

EPA requested that these documents use an environmental "mass balance" approach, i.e., that they attempt to account for the sources, sinks, and receptors for each pollutant as it moves through the environment. These documents were also to explore the technology and means for controlling the pollutant and the resulting costs and benefits.

Within the NRC, responsibility for the study was assigned to the Environmental Studies Board of the Commission on Natural Resources. In agreeing to undertake these studies, the Environmental Studies Board identified an additional objective for the assessments that were to be made. This was to draw upon the experience gained in these assessments to address the broader methodological problems of how such assessments should be done and how best to use the limited resources of scientific expertise on environmental issues to meet EPA's expanding need for independent, critical, scientific evaluations of environmental pollutants. A Coordinating Committee on Scientific and Technical Assessments of Environmental Pollutants (STAEP) was then appointed to oversee panels that would undertake these studies and to make recommendations regarding the best methodology for producing such assessments.

EPA and NRC agreed upon two initial assessments, one of nitrates and the other of a group of compounds, the nonfluorinated halomethanes. These pollutants were chosen because they pose particularly complicated assessment problems, given the current scientific and technical knowledge about them.

In September 1976, EPA requested additional assessment reports under the STAEP study program. The purpose of the request was to help meet the requirements of a consent agreement between EPA and three litigant groups in which EPA agreed to determine the health and ecological effects of a number of chemical compounds. The two resulting additional STAEP studies are the present report on Kepone, mirex, and hexachlorocyclopentadiene, and a report on mercury issued in 1977.

This report differs from the initial STAEP studies in some respects. Firstly, the period of study was shorter and, consequently, the scope is narrower: whereas the other studies delineate modes of transport, transformation, and routes of exposure in addition to assessing the ecological and health effects of a given pollutant, this study focuses primarily on effects, and only considers those aspects of transport, transformation, and routes of exposure that are critical to specific organisms and/or systems. Secondly, this study does not address the costs and benefits of various control technologies. Thirdly, this report is based primarily on a literature survey that was prepared for EPA by another contractor rather than on a survey undertaken by the Panel itself.

The STAEP Panel on Kepone/Mirex/Hexachloro-cyclopentadiene was appointed in December 1976 and was charged with providing EPA with a critical assessment of the available scientific and technical knowledge of the ecological and human health effects of Kepone, mirex, and hexachlorocyclopentadiene as environmental pollutants. The Panel was also to identify areas in which scientific evidence is uncertain or too inconclusive to arrive at an assessment, and to suggest areas where additional research is needed to provide a sounder basis for future regulatory action by EPA.

The principal document used in the effort is one of a series of "REEPs," or Reports on Environmental Effects of Pollutants, which have been developed for EPA by various contractors to assist the Agency in pursuing its regulatory options. The REEP used by this Panel was The Environmental Impact of Mirex and Kepone, prepared by Battelle Columbus Laboratories in 1978 for the EPA Office of Health and Ecological Effects, Office of Research and Development. Another document used as a principal reference for this report was Human Population Exposures to Mirex and Kepone, prepared by B.E. Suta of the Stanford Research Institute in 1977. Panel members evaluated the portions of these documents that relate to their respective disciplines and they combined their expertise and judgment to interpret and assess the implications of the information for the present report. The Panel also solicited information from other sources, including government and international organizations; environmental groups; professional, scientific, and trade organizations; and numerous individuals. The lack of information on hexachlorocyclopentadiene and the proprietary nature of data on Kepone and mirex limited the amount of information available for consideration by the Panel. Consequently, considerable time and effort were spent in seeking additional information from various government agencies and pesticide manufacturers.

The Panel appreciates the contributions and information provided by the individuals and organizations listed in the Appendix. We are especially grateful to Dr. Ian C.T. Nisbet, Chairman of the STAEP Coordinating Committee, who served as liaison to this Panel. We are also grateful to Dr. Charles C. Brown of the National Cancer Institute for his collaboration in the preparation of the section on Quantitative Estimates of Risk.

The Panel would also like to thank the professional staff of the National Academy of Sciences/National Research Council. This support was chiefly provided by Lawrence C. Wallace, Project Staff Officer, who was responsible for coordinating the work of the Panel with information from EPA, the STAEP Coordinating Committee, and other organizations and individuals. He was also largely responsible for the completion of this report. The Panel further expresses its appreciation and thanks to Connie Reges, STAEP Project Secretary, for her contributions throughout the study; to Judith Cummings, Patricia Marshall, Anne Norman, and Philippa Shepherd who provided editorial assistance; and to the staff of the Manuscript Processing Unit.

Robert J. Livingston
Chairman
STAEP Panel on Kepone/Mirex/Hexachlorocyclopentadiene

FINDINGS AND RESEARCH NEEDS

This report's assessment of the scientific and technical
knowledge about the effects of Kepone, mirex, and hex as
environmental pollutants is based primarily on two
literature surveys prepared for EPA's Office of Research and
Development by Battelle Columbus Laboratories and the
Stanford Research Institute. The principal findings and
research needs arising from the Panel's assessment are
summarized below. Documentation for the findings can be
found in the body of the report as noted parenthetically
after each finding. The list of research needs identifies
areas where more knowledge is needed before a truly
comprehensive assessment of the effects of Kepone, mirex,
and hex can be made.

FINDINGS

General

• According to criteria based on population
dynamics, altered energy transfer and
interpopulation response, and community reaction to
chronic exposure to Kepone, mirex, and hex, data on
which to base an informed assessment of
environmental impact are limited. While the data
we do have suggest that species resistance and
adverse chronic effects are high, with proven
bioconcentration and high chronicity factors for
Kepone and mirex, definitive environmental changes
due to varying levels of exposure remain
undetermined. While there are indications that
widespread use of mirex can lead to altered
population structure in terrestrial systems, with
resurgence or escalation of nontarget pests due to
selective (mirex-induced) mortality of predators,
there are too few data for a uniform prediction of
effect. Since Kepone, mirex, and hex are capable
of bioaccumulation and acute or chronic toxicity,
lack of observed effects in areas where they have
been released is insufficient basis for the
assumption that no impact has occurred (see Chapter
3, sections on Criteria for Assessment, Acute and
Chronic Effects, and Ecological Implications).

• In comparison with other environmental
contaminants such as dieldrin, DDT, and PCBs, the
scientific literature about Kepone, though growing,
is incomplete; the knowledge of residue
distribution and acute or chronic effects of mirex
is substantial; and data on the environmental
significance of hex are scarce. However, despite
the relatively small quantities of Kepone and mirex
that have been released into the environment in the
United States, environmental levels of these
compounds are disproportionately high. Thus, while
the magnitude of the hazards to human health and
the environment posed by these residues cannot be
determined accurately, assuming the worst set of
circumstances, they could be substantial (see
Chapter 3, section on Ecological Implications).

Occurrence, Properties, and Toxicity of Kepone

• Exposures to Kepone that are of primary
environmental concern are the Hopewell, Virginia,
incident of industrial release and contamination of
the James River-Chesapeake Bay system; the use of
Kepone against banana root borers in tropical
areas, which could be a cause for concern in the
United States because a significant portion of this
use occurs in Puerto Rico; and the sporadic
appearance of Kepone in mothers' milk in an area
roughly corresponding to the area sprayed with
mirex for control of the imported fire ant (see
Chapter 1, section on Patterns of Contamination;
and Chapter 2, section on Distribution of
Exposures).

• Kepone is environmentally persistent and
capable of movement through ecosystems; it
undergoes bioconcentration and biomagnification and
has been found to occur in human tissue and milk
(see Chapter 1, section on Patterns of
Contamination; and Chapter 2, section on
Distribution of Exposures).

• Kepone is highly toxic to certain organisms.
Its documented chronic effects on some invertebrate
aquatic species include reduced shell growth, loss
of equilibrium, and impaired reproductive
capability (see Chapter 3, section on Acute and
Chronic Effects).

• While generally less acutely toxic to
terrestrial than to aquatic species, Kepone has had
observed chronic effects on some terrestrial
species, including neurological and reproductive

2

disorders; adverse effects related to energy production and muscular function have also been noted in areas of contamination (see Chapter 3, section on Acute and Chronic Effects).

•   There is very little information about the short- and long-term implications of Kepone distribution in a range of aquatic systems (see Chapter 3, sections on Acute and Chronic Effects and Ecological Implications).

Occurrence, Properties, and Toxicity of Mirex

•   There is a clear pattern of mirex distribution in the South in human adipose tissue corresponding to the use of mirex for control of the imported fire ant (see Chapter 1, section on Patterns of Contamination).

•   Mirex has also been manufactured and sold under the trade name Dechlorane for use as a fire retardant in plastics. Effluents from manufacturing processes have resulted in local contamination of Lake Ontario. The fate of large quantities of Dechlorane remains unknown (see Chapter 1, section on Patterns of Contamination).

•   Mirex is one of the most environmentally stable members of the organochlorine insecticide group; it has a tendency to bioaccumulate and biomagnify, and degrades into a series of potentially hazardous products which apparently include Kepone (see Chapter 1, sections on Properties of the Compounds and Dynamics in the Environment).

•   Mirex shows a range of acute toxicity (from low to high, depending on the species), and has one of the highest chronicity factors observed for any pesticide, largely because of its relatively slow rates of metabolism and excretion, which result in its accumulation in biological systems. These effects, together with the extreme sensitivity to mirex of certain groups such as aquatic crustaceans, suggest that the continued broadcast use of mirex could lead to serious environmental problems, including bioaccumulation and/or alterations of species composition and community function (see Chapter 3, sections on Acute and Chronic Effects and Ecological Implications).

Potential Effects of Kepone and Mirex on Human Health

•   Kepone has had toxic effects on humans, and
mirex would be expected to affect humans at levels
of exposure comparable to those producing toxicity
in experimental mammals (see Chapter 2, sections on
Qualitative Estimates of Risk and Quantitative
Estimates of Risk).

•   Kepone and mirex are carcinogenic to
laboratory rodents, mainly causing liver tumors.
Although their mode of action is unknown, and the
relevance to human risk of the tumors induced under
the conditions studied is not yet established, they
must be presumed to be human cancer hazards (see
Chapter 2, section on Qualitative Estimates of
Risk).

•   The potential magnitude of their hazards,
however, is impossible to predict with certainty
because of deficiencies in present knowledge of the
process of extrapolation from results in
experimental animals to human risk.   Nevertheless,
using the best information available, the projected
risk might not be negligible (see Chapter 2,
section on Quantitative Estimates of Risk).

Occurrence, Properties, and Toxicity of Hex

•   Hex and its metabolites undergo
bioconcentration and biomagnification in simulated
microcosms, thus showing environmental
characteristics associated with some of the better
known organochlorine compounds.   These results must
be viewed as preliminary since there are few
reliable data available (see Chapter 1, section on
Dynamics in the Environment).

•   Hex continues to be used and distributed as a
precursor for other chemicals.   Although tens of
millions of kilograms are produced annually, there
is virtually no information on residue distribution
and chronic effects on various organisms.   Acute
toxicity data are also inadequate for assessment of
effect (see Chapter 1, sections on Patterns of
Contamination and Dynamics in the Environment; and
Chapter 3, section on Acute and Chronic Effects).

•   Despite recent restrictions on the use of
several chlorinated cyclodiene insecticides, such
as aldrin and chlordane, production of hex is
expected to continue at a high level for the
manufacture of flame retardants and of pesticides

for export (see Chapter 1, sections on Patterns of Contamination and Dynamics in the Environment).

RESEARCH NEEDS

The following are identified as areas where more information is desirable; they are not presented in order of priority.

• The impact of Kepone on systems associated with areas of treatment for the banana root borer should be investigated. Investigation should include thorough analysis of residues in soil, runoff into surface waters, residues in fish and domestic animals, and residues in tissues of human residents of the area. Data on the use of Kepone to control the banana root borer are also needed, including the amounts of pesticide sold, acreages of its use, and mode and rate of application.

• Homes and commercial establishments where Kepone has been used for ant and cockroach control should be surveyed to determine the extent to which residues of Kepone persist in dust and in indoor air, and whether such residues pose significant hazards to occupants.

• Surveys of human adipose tissue and mothers' milk for Kepone and mirex should be extended to include subpopulations identified as at high risk of exposure, including consumers of seafood from known contaminated sources and workers who handle Dechlorane.

• More dose-response data on the carcinogenicity of Kepone and mirex are needed, as are mechanistic studies on the experimental carcinogenicity of Kepone and mirex, particularly of mutagenicity and ability to interact directly or indirectly with DNA.

• The most important issue concerning Dechlorane relates to materials flow assessments. Critical questions to be answered include determinations of the products in which Dechlorane is used, how long they are used, whether their use results in direct human exposure, where they are discarded, and the ultimate fate of the Dechlorane they contain. Since Dechlorane has been used by many small formulators in plastics industries, environmental contamination from such activities should be assessed. Factory discharges to aquatic systems are particularly important point sources to

5

examine. Initial work in this area could focus on dumps, incinerators, product-recycling machinery, and effluents released into aquatic systems.

• Basic laboratory and field (residue) information is needed on the potential environmental distribution of hex and its acute and chronic effects on a representative range of terrestrial and aquatic species.

• Development of a reliable analytical methodology for detecting residues of hex is critical to any research devoted to determining its environmental distribution.

• More effort should be devoted to studies of the metabolic aspects of hex since its reactivity suggests the potential for transformation and reaction with other chemicals. Such reactivity, including polymerization, could yield mirex-type compounds, and reaction with olefins could yield chlorinated cyclohexene derivatives.

• Bioassay of the carcinogenicity and mutagenicity of hex is needed, and its toxic effects on fetuses and newborns should be determined.

• Epidemiological studies of workers who have handled hex would be desirable.

• Information on changes at the population and community levels caused by the presence of Kepone, mirex, or hex is virtually nonexistent. Any further extensive use of these compounds should be preceded by a combined residue analysis and field evaluation program to determine short- and long-term effects at these levels of organization. If use is discontinued, field studies in areas of contamination should be carried out to determine recovery.

OVERVIEW

The cyclodiene insecticides Kepone and mirex are
synthesized from hexachlorocyclopentadiene (hex). Both
Kepone and mirex are members of the organochlorine group and
are generally similar with respect to their chemical
characteristics, biological activity, and behavior in the
environment. They are also often characterized by general
environmental stability and pronounced bioaccumulation.
Previous studies indicate, however, that there are
significant differences among these three compounds,
including their opportunities for entering the environment,
and their distribution and behavior. For this reason, it is
important that Kepone, mirex, and hex be evaluated
individually in determining their environmental impact and
effects on human health. Each of the compounds must be
studied within the context of its individual patterns of
use, exposures of nontarget organisms, and potential acute
and chronic effects at levels commensurate with the
distribution of its residues.

KEPONE

Kepone, the ketone analog of mirex, has been used
primarily outside the United States as a pesticide against
the banana root borer. In the United States, the only
registered uses of Kepone have been as an ant and cockroach
poison. Because most of the Kepone that was synthesized in
this country was for export, the major source of exposure in
the United States is thought to be from point source
contamination at manufacturing sites. Kepone is a
contaminant of mirex and a product of mirex degradation.
Thus exposure to Kepone may also result from the use of
mirex.

Kepone became a subject of attention when expanding
contamination resulted from its manufacture at a plant in
Hopewell, Virginia. Affected personnel of the Hopewell
plant were hospitalized; relatively high residues of Kepone
were found in sediments and aquatic biota of the James River
drainage system, including the Chesapeake Bay; and a large
area was closed to commercial and other fishing to prevent
exposure of a larger population through the ingestion of
contaminated seafood.

The data on hazards to the environment and human health resulting from exposure to Kepone are particularly weak. The compound has been considered a minor pesticide in this country, because it has been used relatively little in the United States; as a result, the literature on Kepone and its residues is relatively sparse.

The Hopewell incident stimulated some EPA research and led to the establishment of recommended action levels for the presence of Kepone in finfish, shellfish, and crabs. Recently, the Agency also analyzed samples of human milk in nine southern states. Approximately 3 percent of the samples tested, in an area roughly corresponding to that where mirex has been used for the control of imported fire ants, contained Kepone at levels ranging from less than 1 ppb to 5.8 ppb. There has been no report of Kepone residues among the general population, where presumably most exposures to the compound would result from the use of ant and cockroach traps. Although no severe effects have been recorded in accidents involving these traps, little is known about the potential for human exposure to Kepone as a result of the compound's translocation by insects and its presence in dust and air in the vicinity of insect traps.

Recent research indicates that Kepone is environmentally persistent and highly toxic to various organisms. It is carcinogenic to laboratory rodents and, therefore, must be presumed to present a cancer hazard to humans. Kepone tends to move through ecosystems, and is subject to relatively efficient mechanisms for bioconcentration and biomagnification as it moves from one trophic level to the next. In certain invertebrate aquatic species, it has been shown to produce such chronic effects as reduced shell growth, loss of equilibrium, and impaired reproductive capability. Chronic effects observed in laboratory mammals at levels of exposure similar to those in the Hopewell, Virginia, area include neurological and reproductive disorders and adverse effects on energy production and muscular function.

Further research is needed to resolve basic questions about the risks associated with the manufacture and use of Kepone. Study of the residues of Kepone in the general population is needed followed by further study of its impact on human health and the environment.

MIREX

Mirex is one of the most stable of the organochlorine pesticides, and has been used widely in the southern United States for control of the imported fire ant. An estimated 74 percent of the mirex used in the United States for nearly 20 years, however, has been used for nonagricultural

purposes on a nationwide basis. Under the trade name
Dechlorane, the compound is widely used as a fire retardant
in plastics.

There has been public controversy over the use of mirex
since it was first used in 1962 in the program for control
of the imported fire ant. Proponents of this use point out
that relatively low concentrations are needed (4.2 g/ha or
1.7 g/acre) and that there is a lack of feasible
alternatives for combating the imported fire ant, which
causes modest to very severe local losses in some
agricultural operations. Opponents stress that mirex has
proved to be highly persistent under various environmental
conditions and has an established tendency to bioaccumulate
and biomagnify in food webs. As a result of the
controversy, the production of mirex for imported fire ant
control in the United States is being phased out. The sole
producer of mirex that was marketed as a pesticide for use
in the United States has turned over its remaining stocks of
the compound to the State of Mississippi, which with the
U.S. Department of Agriculture, controls the use of these
stocks under EPA restrictions that call for application in a
more diluted form than in the past.

Evaluation of the distribution and environmental effects
of mirex residues is complicated by the use of mirex as the
fire retardant Dechlorane. Dechlorane has been produced at
Niagara Falls, New York, and formulated at a number of sites
in the United States and is used in the manufacture of
plastic products that are distributed nationwide. It is
difficult to assess potential exposures to mirex because the
fate of a relatively large quantity of mirex that has been
produced and used as Dechlorane has not been determined.
The only documented accumulations of mirex in humans are in
southern states, where the compound was used as a pesticide.
It has been reported that diffused, low-level contamination
has occurred in sediments of Lake Ontario and in the aquatic
ecosystem of the St. Lawrence River as a result of releases
of Dechlorane from two manufacturing plants. Although there
are no reports of mirex residues in humans other than in the
South, there has been no concerted effort to look for
residues in other areas.

There is a considerable body of information on some of
the environmental characteristics of mirex, even though its
fate in the environment and the associated transfer
mechanisms are not well defined. Mirex is resistant to
degradation and metabolism and has an environmental half-
life of five to twelve years. It tends to accumulate in
terrestrial and aquatic systems and shows evidence of
biomagnification as it moves through these systems. When it
undergoes decomposition, one of its degradation products is
Kepone. Other stable degradation products (e.g., mono- and
dihydromirex) have been reported in environmental samples,

9

but as yet little is known of their distribution and potential effects.

The acute toxicity of mirex ranges from low to high, depending on the species. It has a high potential for chronic toxicity largely because of its ability to accumulate in biological systems. On the basis of results from laboratory studies, it must be considered a potential human carcinogen. The insufficiency of field data on mirex, however, makes it difficult to develop an unequivocal appraisal of its occurrence, activities, and effects in the environment.

HEXACHLOROCYCLOPENTADIENE

Hex is a highly reactive chlorocarbon that readily undergoes Diels-Alder reaction with other unsaturated organic compounds and forms a variety of chlorinated polycyclic compounds. These characteristics have led to its use primarily as a precursor for the preparation of various insecticides (including Kepone and mirex) and fire retardants. It also has a minor use as a herbicide. However, the publicly available data for hex are scanty and inconsistent; in particular, information about its use, residue characteristics, and biological impact on various organisms and systems is uniformly inadequate. Basic questions about the scope and nature of its potential for environmental contamination remain unanswered. One reason for this lack of knowledge has been the use of hex as a precursor in the formulation of Kepone, mirex, and other products. Attention has focused on its reaction products, with the result that its overall impact has been largely overlooked. In 1977, however, incidents of environmental contamination in Kentucky and Michigan aroused concern for its potential hazards.

The possibility of widespread exposure to this compound has continued despite a reduction in the use of cyclodiene pesticides in the United States. Hex continues to be produced in this country as an intermediate both in the manufacture of fire retardants and of pesticides for export. Future trends in the rate of production are difficult to estimate.

In view of the varied and extensive uses of hex, the potential of related cyclodienes for toxic action and environmental persistence, and the inadequacy of data about the potential effects of hex on various organisms, there is the need for both research on and a comprehensive review of the mass balances and environmental significance of this compound. It is particularly important that its environmental persistence, toxicology, and potential effects on human health be studied.

CHAPTER 1

ENVIRONMENTAL DISTRIBUTION, TRANSPORT, AND FATE OF KEPONE, MIREX, AND HEXACHLOROCYCLOPENTADIENE

PROPERTIES OF THE COMPOUNDS

Kepone

Kepone®[1] (GC1189; chlordecone; decachlorooctahydro-1,3,4- metheno-2H-cyclobuta [cd] pentalen-2-one) is the ketone analog of mirex (see Figure 1.1). Like mirex, it has easily defined chemical and physical properties, and saturated, symmetrical molecules. The key physicochemical characteristics of Kepone, mirex, hex, and of three well-studied environmental contaminants, DDT, dieldrin, and γBHC are summarized in Table 1.1.

The presence in Kepone of a carbonyl group in place of two chlorine atoms present in mirex greatly affects Kepone's solubility in water which is 2,000 times that of mirex (see Table 1.1). The water solubility of Kepone suggests that its lipophilic properties and partition behavior are more like those of γBHC, another chlorinated hydrocarbon. Kepone is also more reactive and volatile than mirex. Its thermal decomposition point is about 400°C, compared to about 600°C for mirex and, judging by the behavior of Kepone in gas chromatography, its volatility is closer to that of dieldrin, another pesticide in the organochlorine group, than it is to that of mirex.

Technical preparations of Kepone contain 94.4 percent Kepone, with 0.1 percent hex as a minor contaminant (U.S. EPA 1978).

Mirex

Mirex (GC1283; Dechlorane®[2]; dodecachlorooctahydro-1,3,4- metheno-2H-cyclobuta [cd] pentalene) is a fully chlorinated, cage-structured compound formed by Diels-Alder reaction from hex (see Figure 1.1). Its resistance to heat (decomposition at 650°C) and its low reactivity with acids, bases, and other chemical agents such as ozone and lithium

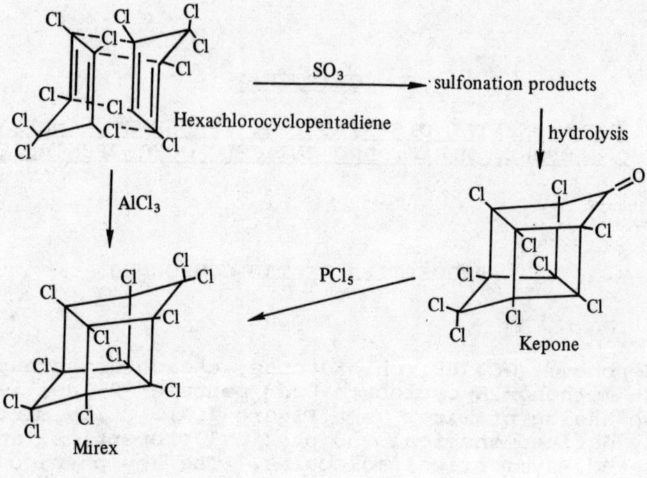

SOURCE: U.S. EPA (1978).

FIGURE 1.1 Synthesis of Kepone and Mirex.

TABLE 1.1 Physicochemical and Environmental Characteristics of Selected Chlorinated Compounds

Compounds	Water Solubility (ppb)	Vapor Pressure (mm Hg)	Estimated Half-Life in Soil (years)
Kepone	2,000	10^{-5}	--
Mirex/Dechlorane	1	6×10^{-6} [b]	>12 [c]
Hex[a]	800	1 [b]	--
DDT	2	1.5×10^{-7}	2.5 [d]
Dieldrin	250	1.8×10^{-7}	2 [d]
γBHC	10,000	9.4×10^{-6}	2 [d]

[a] Hexachlorocyclopentadiene.
[b] At 60°C (Ungnade and McBee 1958); all other vapor pressures at 20-25°C (Matsumura 1975).
[c] Holden (1976).
[d] Extrapolated from data presented in Edwards (1966).

aluminum hydride (Brooks 1974), show it to be a very
unreactive compound.

The physicochemical characteristics of mirex indicate a
potential for stability and suggest that it is persistent in
the environment and may accumulate in biological systems.
The low solubility of mirex in water is comparable to that
of DDT. Since both mirex and DDT are reasonably soluble in
organic solvents, they are likely also to have comparable
partition coefficients and lipophilic properties; in
addition, the volatility of mirex is comparable to that of
DDT. Yet mirex is chemically more stable than either DDT or
dieldrin.

Because of the chemical stability of mirex, no major
degradation product or contaminant other than Kepone appears
to be important at the time that mirex is introduced to the
environment. Technical-grade preparations of mirex consist
of 95.19 percent mirex with 2.58 ppm Kepone as a contaminant
(U.S. EPA 1978).

Hexachlorocyclopentadiene

Hex (C-56; 1,2,3,4,5,5-hexachlorocyclopentadiene) is
widely used as a precursor for the synthesis of Kepone and
mirex (see Figure 1.1), and many other pesticides and
industrial chemicals. Unlike Kepone and mirex, however, hex
is very reactive and volatile (see Table 1.1). Hex is also
nonflammable and its presence is easily detected because of
its pungent odor at relatively high concentrations (0.15 to
0.33 ppm in air [Equitable Environmental Health, Inc.
1976]).

The reactivity of hex comes from its ability to
polymerize even at low temperatures between 20 to 200°C
(U.S. EPA 1978), and to react with olefins and other organic
molecules such as aromatic compounds. Thus, the apparent
disappearance of hex from the environment in routine multi-
residue analyses should not be construed to mean that it is
always degraded to smaller molecules, even though stable
bioaccumulated metabolites of hex have been found (Lu et al.
1975). The apparent loss of hex may result primarily from
volatilization, polymerization, and reactions with other
organic chemicals.

All industrial methods used to prepare hex could induce
the formation of contaminants. The three basic methods of
synthesizing hex are: (1) chlorination of cyclopentadiene;
(2) condensation of small chlorinated hydrocarbons; and (3)
chlorination of n-pentane-isopentane. The first reaction
gives octachlorocyclopentene as an intermediate and
contaminant; the last two reactions could give chlorinated,

14

low molecular weight aliphatic compounds, which may be
difficult to remove by simple distillation.

ANALYTICAL PROBLEMS

Kepone

Kepone has not been sought in any of the routine U.S.
Food and Drug Administration market-basket surveys of
pesticide residues and other pollutants in food. Moreover,
Kepone is not a target of EPA's national pesticide
monitoring program (U.S. EPA 1977a), and routine assays
exclude Kepone in the process of cleanup by Florisil column
and in hexane-acetonitrile partitioning processes.

A satisfactory method for detecting Kepone in tissues,
particularly in bioassays of fish and other aquatic
organisms, was not devised and agreed upon until late in
1976. The late development of this method raises the
question of whether the presence of Kepone in biological
specimens was overlooked or was not accurately quantified.
The limit of detection with current techniques is reported
to be 0.005 ppm in analyses where 400-g samples of banana
peel and pulp are used (U.S. EPA 1977a). This procedure is
said to be applicable for certain other fruits, vegetables,
and milk. Data derived before this degree of sensitivity in
analyses of Kepone residues was available must therefore be
interpreted with caution.

Gas chromatography used in the analysis for residues of
Kepone presents only standard problems of quality control in
processing, that is cleanup and recovery. Kepone, unlike
mirex, is not difficult to identify.

Mirex

Analysts generally experience two major problems in
correctly detecting and identifying the chemical and
physical properties of mirex. The first is the similarity
of the behavior of mirex in gas chromatography to that of
some of the components of PCBs (e.g., Arochlor 1254), which
makes it difficult to determine whether mirex or PCB is
present in the column or whether both are present and
interact.

The second major problem is the long retention time (Rt)
of mirex relative to other similar chlorinated pesticides
(for instance, the Rt for mirex is approximately 5.0 times
that of aldrin [U.S. EPA 1978]). The longer Rt presents a
practical problem in routine multi-residue assays, inasmuch
as compounds appearing at the end of gas chromatograms give
flat peaks with relatively low resolution, and are therefore

15

difficult to identify. Moreover, since practically no other pesticide residues are expected to appear in these areas, the low regions are often excluded from routine surveys. The problem can, however, be overcome if a gas chromatographic condition is set up especially for mirex.

Interference by PCBs in chromatographic analyses is the more serious of the two problems. Since both PCBs, which are also major contaminants of the environment, and mirex are very apolar compounds, they are usually eluted at the beginning of silica gel or Florisil columns, and their separation requires rather careful cleanup treatments of the chromatographic columns. Markin et al. (1972) point out two examples in the literature where these compounds may have been incorrectly identified.

The Panel believes that, despite these problems, most of the methods used for detection of mirex residues are reasonably reliable. Such residues have reflected patterns of use in the South, for example, where mirex was broadcast for a period of nearly 20 years as a pesticide for control of the imported fire ant. There are few documented cases where mirex, including Dechlorane, has been detected in areas where there was no history of application or possible human exposure. The lack of such chemical epidemiology data is convincing enough to indicate that at least the qualitative aspect of identification of mirex residues is sound.

Hexachlorocyclopentadiene

There is no established or accepted method of analysis for hex. Consequently, there are no data on environmental residues of hex or its noncommercially derived products.

PATTERNS OF CONTAMINATION

Kepone

For several years after the development of Kepone in the early 1950s by the Allied Chemical Corporation, production of the compound was limited and sporadic. Demand built up in the late 1960s as the banana root borer began to develop immunity to other pesticides. Production of Kepone in the United States ended in 1975 after the closing of the plant in Hopewell, Virginia, that had been acquired from Allied by another firm. At the time of its closing, this was the only plant in the United States that was known to be manufacturing Kepone.

Between 90 percent (Sterrett and Boss 1977) and 99.2 percent (U.S. EPA 1976d) of the Kepone that was produced in the United States was exported to Latin America, Africa, and Europe, with most of the remainder used in this country as ant and cockroach poison. The patterns of use for the Kepone that was exported are poorly documented, but it is known that one of its principal uses was for control of the banana root borer, primarily in the Caribbean area, including Puerto Rico. According to EPA (1978), most of the Kepone that was exported to a West German firm was converted into other pesticide products, but this has not been confirmed.

No exact data are available on the total amount of Kepone that was produced in the United States for export and for domestic uses, but it is thought that this production totalled about 1,600,000 kg (3,500,000 lb) (Ferguson 1975). Of this amount, the domestic consumption of Kepone has been estimated at from 12,000 to 70,000 kg (U.S. EPA 1976d, Sterrett and Boss 1977). These estimates are probably too high, however, because they represent the difference between the estimated total production and the estimated exports. They do not take into account factory inventories that remained or losses to the environment during production. The existing stocks of Kepone at Allied, which continued to market its Kepone inventories after production of the technical-grade product ended, were small in 1976 (about 250 kg of Kepone were incorporated in a 25-percent Kepone/special mixture), according to an estimate by EPA (1976e). In 1976, EPA issued a notice of intent to cancel the use of Kepone in ant and cockroach traps; Allied concurred in the cancellation of this insecticide, which is no longer being manufactured. However, formulation of existing stocks of Kepone for use in ant and cockroach traps may continue until May 1, 1978.

Since most of the Kepone synthesized in this country was for export, the major source of contamination from this compound in the United States is thought to have been at manufacturing sites for technical-grade Kepone, the primary product used by firms marketing commercial grade formulations for ant and cockroach traps and the form of the product that was exported. Allied and its subsidiaries were the leading producers of technical-grade Kepone (the plant in Hopewell was later sold to another firm), but at various times two other chemical companies supplied this firm with technical-grade Kepone.

Nearly 30 firms in the United States were using Kepone in the manufacture of ant and cockroach traps when this use of the chemical was ended. The formulation of these products probably did not result in major losses to the environment because the amounts of Kepone used were small. Most traps or baits had only 0.125 percent of the active ingredient and usually incorporated some safety device to prevent direct human contact with the insecticide.

17

James River Contamination

It is not possible to estimate the total quantity of Kepone that entered the James River at Hopewell, Virginia, on the basis of data on Kepone manufacture. Calculations by Smith (1976) show that routine losses from washing, volatilization, and other manufacturing processes can account for only a few kilograms of the Kepone that is now known to have been lost from the Hopewell plant from 1971 to 1974. According to EPA (1976d), as much as 45,000 kg of Kepone now lie on the bottom of the James River. This estimate is based on data derived by the Agency from a study of Kepone residues in sediments of the James' drainage system after it became clear that workers at the Hopewell plant had developed health problems and the possibility arose that the James might also be contaminated. The data on Kepone residues in sediments also show that this one pollution point source contaminated the entire river from Hopewell to Newport News, Virginia. The highest concentration of Kepone (over 10 ppm) was found at Bailey's Creek, the site of the original contamination. Beyond Newport News, the presence of Kepone residues was found to be sporadic and diffused, the pattern being controlled by currents and movements of the sediments.

Contamination of soil around the production site was reported by Blanchard (1976). Total Kepone contamination appeared to be about 1,000 kg within a radius of one kilometer around the Hopewell factory.

Use on Banana Crops

The use of Kepone in Puerto Rico as a pesticide against the banana root borer may require special attention by the United States government. The pesticide label directions call for surface application of 4.2 kg/ha (3.75 lbs/acre) of active ingredient every six months. At this rate, the Panel has calculated that use of the pesticide will result in about a 100 ppm level of Kepone residues in the top 3 cm of soil after each application. In one test reported by United Fruit Company (1969), a single application of Kepone at 6.73 kg/ha (6 lbs/acre) active ingredient led to residues of Kepone in soil from 15 to 25 ppm that persisted for at least six months.

From what is known about the slow dissipation rate of Kepone in the environment, its level in banana fields in Puerto Rico could be quite high. Unfortunately, no data are available on the levels of Kepone residues in soil, chickens, grasses, and other components of the environment in Puerto Rico. Consequently, no calculations have been made concerning impact on humans and the environment.

Use in Insect Traps

In the United States, the only registered uses of Kepone
have been for control of ants and cockroaches in domestic
and commercial establishments. If environmental
contamination had resulted from these uses, it seems likely
that Kepone residues would have been found in the general
population of the entire United States. There have not been
any reports of such residues, however, in human tissues in
the general public, although there are data on Kepone
residues in mothers' milk in the southern states (U.S.EPA
1976b). EPA collected 298 samples of mothers' milk in nine
southern states in 1976. Nine of the samples collected from
three states--Alabama, Georgia, and North Carolina--showed
Kepone levels ranging from less than 1 ppb to 5.8 ppb. No
residues were found in milk samples from Arkansas, Florida,
Louisiana, Mississippi, South Carolina, and Texas. Possible
sources of exposure include the spraying of mirex for
control of fire ants and the subsequent degradation of the
mirex into Kepone (Carlson et al. 1976), and direct exposure
to ant or cockroach traps. However, most reported incidents
of human exposure to Kepone through contact with insect
traps involve young children.

According to the U.S. Department of Agriculture (1977),
56 incidents of human exposure to Kepone have been reported.
Children under five years of age were exposed in 52 of the
incidents; two incidents involved adults; the other two
incidents involved unspecified ages. All but nine of the
young children were exposed to Kepone at home, primarily by
control devices for ants and cockroaches.

Mirex

It is believed that all technical-grade mirex in the
United States has either been manufactured or distributed by
the Hooker Chemicals and Plastics Corporation. Data on the
sales of mirex/Dechlorane by Hooker have been published by
the U.S. EPA (1978) (see Table 1.2) According to these
data, from 1959 through 1975 about 400,000 kg of mirex (26
percent of the total) were produced for use as an
insecticide and about 1,125,000 kg (74 percent of the total)
were sold as Dechlorane. At the end of 1976, the Hooker
plant had 146,000 kg of mirex inventory remaining at the
site, which would have to be sold as Dechlorane or be
exported, because the use of mirex as a pesticide in the
United States is now limited to existing stocks in a plant
in Mississippi. This inventory, plus the estimated sales of
mirex and Dechlorane from 1959 through 1975, indicate a
mirex production totalling roughly 1,671,000 kg. The
purposes for which mirex has been employed suggest that much
of this production has been used in the United States. Use

19

TABLE 1.2 Domestic Sales of Mirex by Hooker Chemicals and Plastics Corporation[a]

Year	Pesticide Use[b]	Nonagricultural Use[c]	Total
1959	--	50	50
1960	--	800	800
1961	90	4,000	4,090
1962	5,400	35,900	41,300
1963	11,600	129,200	140,800
1964	12,700	210,000	222,700
1965	24,500	245,700	270,200
1966	31,300	150,200	181,500
1967	32,900	72,300	105,200
1968	28,100	126,300	154,400
1969	46,300	32,800	79,100
1970	26,300	20,700	47,000
1971	13,600	37,200	50,800
1972	61,500	54,400	115,900
1973	51,300	--	51,300
1974	36,300	1,300	37,600
1975	18,200	4,600	22,800
Total	400,090	1,125,405	1,525,540
Percent	26.2	73.8	100.0

[a] All amounts given in kg.
[b] Used primarily as insecticide for control of imported fire ant.
[c] Used primarily as flame retardant.

SOURCE: U.S. EPA (1978).

estimates are available, however, only for the mirex that
was sold as a pesticide in the United States.

Use for Control of Imported Fire Ants

An estimated 250,000 kg of mirex were used in this
country for control of imported fire ants from 1962 to 1975
(Markin et al. 1972). It has been estimated that at the
standard application rate (mirex 4X bait = 4.2 g/ha of
mirex) the mirex content in soil should range from 4 ppb
(Mirex Advisory Committee 1972) to 5 ppb (Markin et al.
1972) in a standard three-inch soil sample. This magnitude
of soil contamination, several parts per billion, has been
found in areas sprayed with multiple applications of mirex
in the South (U.S. EPA 1978). The states with a history of
the heaviest mirex usage--Louisiana, Mississippi, and
Georgia--also show the largest number of human tissue
samples with mirex residues, as shown in Table 1.3.

In brief, the use of mirex for control of the imported
fire ant is clearly documentable in terms of its history,
locations, and patterns of use; the total quantities
involved are reasonably accounted for. Even though mirex
has been used extensively as a low-volume pesticide, its
residues are clearly detectable and geographically limited
to those areas where it was actually sprayed for control of
fire ants.

The area of fire ant control accounts for 80 percent of
the area where toxaphene was used (Guyer et al. 1971). The
amount of mirex used in the same area amounts to roughly one
400th of the amount of toxaphene used. The most conspicuous
difference in these two cases, however, is that mirex
residues clearly appear in human adipose tissues according
to the geographical distribution of the treated area as
shown in Table 1.3, while no consistent residues of
toxaphene are found in human tissues in the general public
in the same or any other areas of the United States (Hayes
1975, Kutz et al. 1976, Yobs 1971). The above would imply
that mirex has an extremely long environmental half-life, a
fact documented in the literature (Carlson et al. 1976,
Holden 1976); that the mirex pathway to humans is more
direct; that its turnover in humans is slow; or combinations
of all these factors.

Contamination of Lake Ontario

Mirex residues have also been reported at manufacturing
sites. According to reports on residues in Lake Ontario,
there are two major sources of contamination of the lake:
the Niagara River, where the Hooker plant is located, and
the Oswego River, where mirex contamination was traced to an

21

TABLE 1.3 Incidence of Mirex Residues in Human Adipose Tissues
in Eight Southern States

State	Number of Samples Examined	% Positive Samples	Acres Sprayed in 1962-1973 (million acres)
Louisiana	47	40	13.1
Mississippi	28	32	17.3
Georgia	51	24	45.8
Alabama	27	11	4.4
South Carolina	17	6	7.8
Florida	53	6	8.5
Texas	52	0	2.1
North Carolina	9	0	0.9

SOURCE: U.S. EPA (1976c).

Armstrong Cork Company plant that handled Dechlorane as a
fire retardant (U.S. EPA 1978, Task Force on Mirex 1977).
It is difficult to determine the total amount of mirex
released into Lake Ontario from these plants. Holdrinet et
al. (in press), using estimates from inventory losses,
reported that 450 kg of mirex were released into Lake
Ontario within the last 15 years by the Armstrong Cork
plant. The input from the Hooker plant was much more
difficult to assess. Calculations based on estimates of
mirex losses from washing, volatilization, and other
activities at the plant show only insignificant losses,
i.e., 9 kg/year (Task Force on Mirex 1977). The amount of
mirex lost from these plants can best be estimated from data
on residues in lake sediments. These data show that nearly
700 kg of mirex have been estimated to be present in
sediments of Lake Ontario (Table 1.4). The pattern of
contamination in these sediments (Figure 1.2) clearly
indicates that point sources are responsible for the
contamination of the lake.

Dechlorane Contamination

There is little information on the fate of Dechlorane in
the environment, although this product has been widely used
generally as a fire retardant in plastic products in many
different locations. Rough estimates based on Hooker sales
data indicate that nearly three times as much Dechlorane as
mirex has been marketed by this firm. The name Dechlorane
has been used for different products that bear many
similarities to $C_{10}Cl_{12}$ (the empirical formula for mirex).
The proportion of Dechlorane added to plastics is high:
about 25 percent of plastic resins (Equitable Environmental
Health, Inc. 1976). There appears to be no economic
stimulus to recover and recycle Dechlorane after it has been
incorporated in other products.

The patterns of use for Dechlorane indicate that its
residues should be widely distributed throughout the United
States (not just in the South, as was the mirex used in fire
ant control) and that Dechlorane residues should be
concentrated in urban environments, where most plastic
materials are used. In addition, a number of pollution
point sources might be indicated because of the use of
Dechlorane by both large and small manufacturers and
formulators throughout the country. As far as we can judge
from the data to date, however, there is no widespread mirex
contamination of urban environments as a result of
Dechlorane use. The study of mirex residues in Lake
Ontario, for example, does not reveal major contamination of
the northern shore of the lake, where the cities of Toronto
and Hamilton are located. A comparison of the data on mirex
residues in herring gull eggs in the Great Lakes region
(Table 1.5) indicates that those from Lake Ontario have

23

TABLE 1.4 Estimated Residues of Mirex-Dechlorane in Lake Ontario Sediments

Occurrence	Estimates of Mirex Lost at Site (kg)[a]	Surface Area (km^2)	Mean Residues (ppb dried sediments)
Niagara Anomaly	366	2,349	10.0
Oswego Anomaly	224	1,967	7.3
Other Samples	98	1,117	5.6

[a]Calculated from samples of sediment 3 cm deep.

SOURCE: Holdrinet et al. (In press).

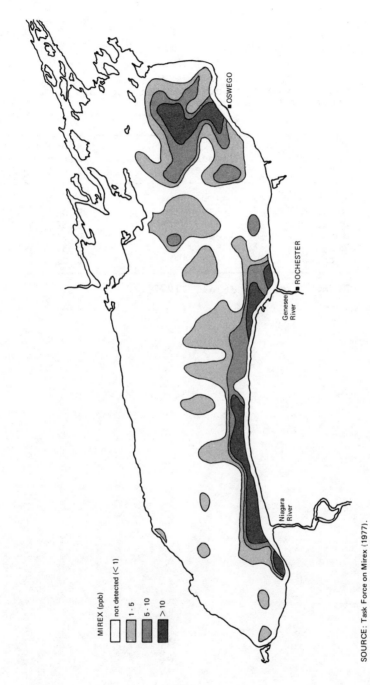

MIREX (ppb)

☐ not detected (< 1)

■ 1 - 5

■ 5 - 10

■ > 10

■ OSWEGO

■ ROCHESTER

Genesee
River

Niagara
River

SOURCE: Task Force on Mirex. (1977).

FIGURE 1.2 Mirex concentration in Lake Ontario sediments.

25

Table 1.5 Mirex Levels in Herring Gull Eggs of the
Great Lakes Area

Lakes	Number of Samples	Average Mirex Level (ppm)
Ontario	39	5.06
Superior	39	0.66
Huron	40	0.56
Erie	42	0.31
Michigan	15	0.01

SOURCE: Gilman et al. (1977)

outstanding levels of mirex, though the levels in eggs from other lakes are also rather high. Mirex has also been detected in several species of fish from the Great Lakes. The highest level reported, 1.39 ppm, was found in an American eel from Lake Ontario (Task Force on Mirex 1977).

While this Panel has found no evidence of significant widespread contamination as a result of the use of Dechlorane, it believes that the need exists for a uniform survey of mirex residues in the environment before a definite conclusion can be made. One reason for the apparent lack of mirex (Dechlorane) residues in areas other than the southern states may be that plastics tightly trap or bind the molecules of this compound and inhibit or prohibit their release into the environment. This particular point has never been established, however, despite its implications for human and environmental health.

Hexachlorocyclopentadiene

The total production of hex in the United States is high compared to that of Kepone and mirex. According to Whetstone (1964), about 22.5 million kg of hex were produced in 1962. Lu et al. (1975) also estimate that the same amount was produced in 1972, and a report prepared for Hooker cites an annual production figure of about 22.7 million kg/yr (Equitable Environmental Health, Inc. 1976). On the basis of these reports, the total production of hex in the 13 year period from 1962 to 1975 is estimated to be about 300 million kg.

The major uses of hex have been in the preparation of insecticides. Minor uses include the preparation of plant-growth regulators, bactericides, fungicides, weed eradicators, high-pressure lubricants, rust inhibitors, rot-resistant additives in plywood, adhesives for rubber and plastics, and catalyst activators (U.S. EPA 1978). These uses, however, are not expected to result in significant levels of hex residues in humans or the environment, except factory workers who may be directly exposed at work.

According to Lawless et al. (1972), the total amount of hex used to produce seven major cyclodiene insecticides was 19.9 million kg, a sizable part of the hex production of about 22 million kg/yr. Amounts of hex used to produce each of three primary derivatives are as follows: chlordane, 11.34 million kg/yr; aldrin, 4.5 million kg/yr; and heptachlor, 2.7 million kg/yr. Because use of these insecticides has probably declined in the United States, one would imagine that hex production might have been reduced. However, currently there is no documentation of this. The reason may be that hex has had continued use as a raw

27

production of fire retardants and of export.

in estimating the environmental hazards of nsurmountable at present. First, there are on residues in the environment; second, there is no way to estimate the amount of free hex present in such commodities as plastic products; third, there is no accepted analytical method to detect hex residues; and finally, the forms of hex reaction products that are stable enough to leave terminal residues in the environment and that may be toxic to biological systems are largely unknown, although they may result in critical impacts on human health and the environment.

An important aspect of hex that should be noted is that it is chemically volatile. Thus, initial study of the impact of hex residues on human health and the environment should focus on residues in the atmosphere.

DYNAMICS IN THE ENVIRONMENT

The environmental distribution and significance of pesticides have been reviewed by various authors (Livingston 1977). There are certain general characteristics among organochlorines that may vary considerably from one compound to another. Many such formulations are characterized by resistance to degradation, complex decomposition mechanisms, environmental stability and persistence, bioconcentration, and biomagnification. Often, these chemical properties depend upon a compound's solubility. While the individualized, compound-specific differences in the behavior of organochlorines indicate the importance of making a detailed evaluation of each compound, the proven environmental stability of this group indicates that their common characteristics should be included in any evaluation of their dynamic relationships in natural systems.

Kepone

Kepone is relatively insoluble in freshwater and in seawater. It leaches readily through various soils, although some studies indicate that under certain conditions it can be stable and persistent in soils (U.S. EPA 1978). It is toxic to certain microorganisms. Aquatic plant and animal species can be highly efficient in accumulating Kepone, and it is known that a large Kepone reserve can be found in the flesh of fish (Suta 1977). The ability of different species to concentrate Kepone varies considerably, however, as a consequence of differences in depuration rates, which can be high in such organisms as oysters and low in some fishes. In general, Kepone is susceptible to transfer from particulate or food-web processes to higher trophic levels with relatively efficient mechanisms for biological magnification, including concentration in humans (U.S. EPA 1978).

Because Kepone has been considered a minor pesticide, little literature exists about the distribution of its residues. Most information on the mechanisms for environmental transfer of Kepone, for example, comes from the incident at Hopewell, Virginia, which represents a single point source and a unique exposure. This incident has made it clear that Kepone is environmentally persistent and has chemical characteristics that tend to follow the behavior of several other organochlorine compounds such as DDT and dieldrin (U.S. EPA 1978).

According to results disclosed at the Fourth Biennial International Estuarine Research Conference (Mount Pocono, Pennsylvania, 1977), Kepone has dispersed through the James River estuarine system via turbulent mixing, river inflow, and tidal fluctuations. There were indications that the deposition rate contributed to the distribution of sediment contamination. Garnas et al. (1977), using static and continuous-flow estuarine microcosms, showed that Kepone desorbs from sediments taken from salt marshes and the James River, and that such processes remain independent of prevailing temperature and salinity. O'Connor and Farley (1977) indicated that the distribution of Kepone in the James River system was affected by physicochemical and biological mechanisms. These included hydrodynamic phenomena (solution, suspension, movement), adsorption and desorption from suspended and bed solids (including settling processes and resuspension), and assimilation-excretion routes through food webs. Bender et al. (1977) found variable Kepone residues in Chesapeake Bay biota, and that such distribution depended on life-history patterns of individual species. Bioconcentration by finfish was species-specific, with considerable dependence on the residence time of migratory fishes. Residues generally declined with distance from the source. Bahner and Nimmo (1977), in a series of experiments, found that Kepone was concentrated by various organisms with bioconcentration factors as high as 13,000 times those under chronic conditions. There were indications that rapid uptake from food and water, relatively slow depuration in certain species, and high water solubility led to the transfer of Kepone through food webs. They postulated that this could pose a threat to consumers.

Mirex

There is a considerable body of information about the behavior of mirex in environmental systems (Collins et al. 1974). In addition to the control of the imported fire ant, mirex also has been used to control the pineapple mealy bug in Hawaii.

Mirex can be concentrated in fishes directly from sediments (Kobylinski and Livingston 1975), water, or food. While photodecomposition products (enhanced by interaction with aliphatic amines) can occur and are presently being used to enhance decomposition in field use, the toxicity of the resulting monohydro, dihydro, and trihydro degradation products remains unknown. In addition, Cripe and Livingston (1977) found that certain photodecomposition products accumulated on bait particles leached by seawater and that organisms in a simulated marsh concentrated one of these compounds in a manner similar to mirex itself. Decomposition products must therefore be included in any evaluation of the "disappearance" of the parent compound.

Mirex residues are quite persistent in various species. The resistance of mirex to degradation and metabolism leads to environmental stability and biomagnification through terrestrial (including the human web) and aquatic systems (U.S. EPA 1978). However, the fate of mirex in the environment and the associated transfer mechanisms have not been well defined. The situation is further complicated by an inability to account for almost half the mirex sold from 1962 to 1973 and, in some cases, the mixing of usage data for flame retardant and fire ant control programs.

Like Kepone, mirex is mobile but, because of its solubility characteristics, it is not readily transported as a dissolved substance in water and probably moves through the environment dissolved in aliphatic materials and/or adsorbed to particulate matter (U.S. EPA 1978). Because of its mode of application, atmospheric contamination and dissemination are unlikely. Extensive residue surveys indicate that various factors are instrumental in the distribution of mirex, including: proximity to treated area, rate of decomposition, rainfall patterns, surface runoff, duration of exposure, seasonal population movements, avoidance behavior, trophic relationships, and other habitat considerations (Suta 1977). Like Kepone, mirex thus possesses chemical characteristics that lead to concentration in nontarget terrestrial and aquatic organisms.

Hexachlorocyclopentadiene

The use of hex for the manufacture of various
insecticides and fire retardant materials will evidently
continue for the immediate future (U.S. EPA 1978). Some
assessment of the environmental movement of this compound is
therefore necessary. Such an evaluation of hex should
include studies of exposure to its by-products and residues,
its potential release as a contaminant of various technical-
grade products, and the environmental characteristics of
fire retardants and hex-based insecticides and their
metabolic by-products.

Few scientific data exist about the movement of hex
through terrestrial and aquatic systems (U.S. EPA 1978). It
is known that the compound is relatively insoluble in water
([0.805 ppm] U.S. EPA 1978), and that it tends to be stable
in the environment and breaks down slowly. Bacteria appear
to provide the most common mechanism for degrading hex,
although some findings indicate that hex is not only
resistant to microbial action but may be toxic to organisms
that are capable of detoxifying it. Lu et al. (1975), using
model terrestrial and aquatic ecosystems, showed that hex
has considerable stability in the environment. It is
concentrated by various organisms and the concentration is
increased as it moves through food chains (e.g., snails,
mosquitoes, and fish concentrate hex 929, 1,634, and 448
times, respectively [Lu et al. 1975]). However, there are
no other available studies and, although there are early
indications that hex acts in a fashion similar to certain
other organochlorines, such as DDT, in the terrestrial and
aquatic environment, there is little information about the
behavior of the compound under field conditions.

NOTES

1   Kepone® indicates a registered trademark. In the rest
of the report the product is referred to as Kepone.

2   Dechlorane® indicates a registered trademark. In the
rest of the report the product is referred to as
Dechlorane.

CHAPTER 2

EXPOSURES TO HUMANS AND ESTIMATES OF CONSEQUENT RISKS

DISTRIBUTION OF EXPOSURES

The potential for human exposure to Kepone and mirex has been assessed by the Stanford Research Institute (Suta 1977) on the basis of a detailed, review of the occurrence of residues in environmental samples. The study's overall estimates of human exposures are summarized in Tables 2.1 and 2.2. According to these estimates, the most important human exposures to Kepone and mirex (other than exposures of very small numbers of persons to locally high concentrations near manufacturing sites and polluted rivers) are the following:

Kepone: From consumption of mothers' milk in the southern United States; average daily intake of the order of 1 microgram, involving some 3,000 breast-fed infants at any one time.

--From consumption of seafood from Chesapeake Bay; average daily intake of the order of 1 microgram, involving 5 to 10 million people.

Mirex: From consumption of seafood from Lake Ontario, the St. Lawrence estuary, and the southern United States; average daily intake of the order of a few tenths of a microgram, involving up to 5 million people.

--From consumption of game birds and mammals in the southern United States; average daily intake of the order of a few tenths of a microgram, involving roughly 9 million people.

Hex: No estimates of exposure were made by Suta (1977); this reflects the absence of any specific data on the distribution of hex in the environment.

The Panel accepts the estimates in Tables 2.1 and 2.2 as a reasonable interpretation of the available data; the data themselves, however, are fragmentary and incomplete. The Panel has identified several problem areas in which human

32

TABLE 2.1 Human Population Exposures to Environmental Kepone

Source	Average Environmental Concentrations	Calculated Average Adult Daily Exposure	At-Risk Population
Food			
James River seafood[a]	0.4-2.0 ppm, finfish	1.1μg[b]	NE[d]
	0.3-3.0 ppm, crabs	8.5μg[c]	
	0.1-0.2 ppm, oysters		
Chesapeake Bay seafood	0.01-0.08 ppm, finfish	0.3μg[b]	5-10 million
	0.008-0.05 ppm, oysters	3.5μg[e]	
	0.10-0.26 ppm, crabs		
Select East Coast Atlantic Ocean Fish	0.01-0.04 ppm[f]	<0.27μg[g]	>570,000
Spring Creek fish[h]	0.025-0.23 ppm	0.02-0.20μg[i]	very few
		0.14-1.33μg[j]	
Mothers' milk	<3 ppb	0.6-1.9μg	3,300
Atmospheric			
Basic product manufacturing neighborhoods	<50μg/m^3[k]	<750μg[k]	115,000
Indoor ant bait use	<0.3ng/m^3[l]	<4.5ng[l]	<6-12 million/yr combined
Indoor ant trap use	<9ng/m^3[l]	<135ng[l]	
Drinking Water			
Lower James River	<0.1-10 ppb	<0.15-15μg	very few[m]
Tobacco			
No concentrations reported	1 ppt[n]	0.004ng[n]	<35% of U.S. adults

[a] The James River estimates assume that the river is open with no restrictions.
[b] Based on consumption on a species basis.
[c] Based on eating only seafood taken from the James.
[d] Not estimated because the river is currently closed to the taking of many species.
[e] Based on eating only seafood taken from the Bay.
[f] Mainly bluefish.
[g] Based on consuming only bluefish as the entire finfish component of the diet.
[h] Catch taken near the Nease plant.
[i] Based on freshwater finfish consumption.
[j] Based on all finfish consumption.
[k] These exposures no longer exist because the basic product is no longer manufactured.
[l] Due to volatilization, concentrations could be higher due to suspension of detached bait particles.
[m] Although these concentrations have been reported for the James, they are not applicable to any municipal water supply.
[n] The estimate is based on plant root uptake data; no concentration data have been reported.

SOURCE: Suta (1977).

TABLE 2.2 Human Population Exposures to Environmental Mirex

Source	Average Environmental Concentrations	Calculated Average Adult Daily Exposure	At-Risk Population
Food			
Lake Ontario seafood	<0.01-0.2 ppm	0.05 μg[a] <0.34 μg[b]	<1 million
St. Lawrence seafood	0.02-0.10 ppm	0.06 μg[a] 0.39 μg[b]	<100 thousand
Southeastern seafood[c]	0.01-0.03 ppm	0.02 μg[a] <0.13 μg[b]	<4.5 million
Spring Creek fish[d]	0.02-1.00 ppm	0.02-0.09 μg[a] 0.12-5.80 μg[b]	very few
Southeastern wild game	<0.06 ppm	<12 μg[e] <0.1 μg[j]	<9 million
Atmospheric			
Basic product manufacturing neighborhoods	NE[f]	NE[f]	130,000
Areas of mirex bait application	<0.006 ng/m^3[g]	<0.09 ng[g]	<300,000
Drinking Water			
No contaminated supplies found	<0.1 ppb[h]	<0.15 μg[h]	NE[f]
Tobacco			
No concentrations reported	<1 ppt[i]	0.004 ng[i]	<35% of U.S. adults

[a] Based on freshwater finfish consumption.
[b] Based on all finfish consumption.
[c] Game or fish taken from areas in which mirex bait has been applied.
[d] Catch taken near the Nease plant.
[e] Considered an extreme overestimate; based on wild game fulfilling the requirement for all meat in the diet.
[f] Not estimated; exposures no longer exist.
[g] Considered to be an upper limit.
[h] This is the lower limit of detection for most sampling; no water supplies were found at this concentration.
[i] An estimate based on plant root uptake data; no concentration data have been reported.
[j] Average per capita consumtion.

SOURCE: Suta (1977).

exposures might prove to be greater than those indicated in Tables 2.1 and 2.2.

Kepone from Ant Traps

In the United States, Kepone has generally been used indoors in the form of traps and baits against ants and cockroaches. Direct calculation of the concentrations of Kepone in the atmosphere of buildings resulting from this use involves a number of uncertain assumptions (Suta 1977); better estimates of potential human exposure may be obtained by materials accounting methods. Kepone introduced into a building in an insect trap is translocated by the insects themselves, and after the insects' death is dispersed through the building as vapor, on dust particles, or attached to surfaces. Removal from the building can occur through disposal of the trap, through ventilation, or through sweeping and disposal of dust. Unless the trap is discarded prematurely, all these processes are likely to be slow, affording prolonged opportunity for exposure of the occupants of the building. In a building in which three traps are used annually, as much as 22.5 mg of Kepone might be dispersed each year (Suta 1977). If only 5 percent of this quantity were inhaled or ingested by the occupants, this would correspond to total absorption of 3 µg per day. Since other pesticides such as chlordane are known to be concentrated in indoor air and home dust (Starr et al. 1974, Tessari and Spencer 1971) this is potentially a major route of human exposure to Kepone.

Kepone and Mirex in Human Milk

Kepone has been detected in only 9 of 298 samples of human milk analysed in the United States, and mirex has not been detected at all (Suta 1977). However, the detection limits were fairly high--1 ppb and 30 ppb respectively--and Kepone was sought only in samples from the southern United States. Extensive studies of other chlorinated hydrocarbons have shown that they are stored in human fat, and secreted in milk at concentrations similar (on a lipid basis) to those in the tissues of the mother (see Table 2.3). Accordingly, the daily intake of these compounds by breast-fed infants is typically about the same as that of their mothers--i.e., it is 5 to 30 times higher on a µg/kg basis (Table 2.3). In view of their similar properties (stability and lipophilicity), Kepone and mirex would be expected to behave similarly.

Hence, for _Kepone_, breast-fed infants whose mothers consume seafood from Chesapeake Bay would be expected to ingest roughly 1 µg per day (close to the detection limit in milk). For _mirex_, the occurrence in human adipose tissues

in individuals from the southern United States would be
expected to lead to corresponding levels in human milk.
Estimating a mean level of about 0.1 ppm in maternal adipose
tissue (U.S. EPA 1978), this would be expected to lead to
ingestion of roughly 2 μg per day by breast-fed infants, by
analogy with trans-nonachlor (see Table 2.3).

Kepone in Banana-growing Areas

Most Kepone has been used for soil treatment of banana
plantations in Central America and the Caribbean area,
including Puerto Rico. In one field test, application of
Kepone to soil at 6.73 kg/ha (active ingredient) led to soil
residues in the range of 15 to 25 ppm, persisting for at
least six months (United Fruit Company 1969).

The potential for human exposure thus exists, through
consumption both of livestock which may forage in treated
areas, and of fish in surface waters polluted by runoff.
Although specific data for areas treated with Kepone are
lacking, analogous data for dieldrin show the potential for
contamination by erosion of soil from treated fields. In a
study in Ontario, streams draining agricultural land in
which dieldrin residues in soil averaged about 0.16 ppm
contained dieldrin at concentrations around 1 ng/l, and fish
in these streams contained dieldrin at concentrations in the
range of 0.003 to 0.23 ppm with a mean concentration of
about 0.04 ppm (Frank et al. 1974). Similar data have been
obtained in Iowa in which dieldrin residues in the range of
0.01 to 1.6 ppm in fish have been associated with runoff
from fields containing soil dieldrin residues in the range
of 0.1 to 0.5 ppm (Morris and Johnson 1971, Kellogg and
Bulkley 1976). Assuming that Kepone would behave similarly
to dieldrin, runoff from soils containing 20 ppm Kepone
would lead to residues in fish of the order of 10 ppm--two
orders of magnitude higher than those in fish in the
northeastern United States.

QUALITATIVE ESTIMATES OF RISK

Kepone

Kepone is lethal as a single dose to rats, rabbits, and
dogs (U.S. EPA 1978). It is not metabolized in mice (Huber
1965), and in keeping with a lack of metabolism for
toxicity, the sex of the animal makes no significant
difference to toxicity. Thus, it is likely that it would be
similarly toxic to all mammals including humans.

Subacute toxic effects have included growth depression,
tremors, hepatomegaly, and reproductive failure in mice
(Huber 1965) and rats (United Fruit Company 1969). In male

36

TABLE 2.3 Intake of Chlorinated Hydrocarbons by Breast-fed Infants Compared to that of the Average Adult in the U.S. Population

Chemical	Estimated Daily Intake by Average Adult (µg)[a]	Mean Level in Adipose Tissue Lipid of Adults (ppm)[b]	Mean Level in Human Milk Lipids (ppm)[c]	Estimated Daily Intake by Breast-fed Infant (µg)[d]	Ratio of Intakes, Infant: Mother
P,P'-DDE	5-28	5-7	2.3	87	3-18:1
PCBs	9	ca. 1.3	1.8	68	7.5:1
Dieldrin	8	0.24	0.12	4.6	0.6:1
Heptachlor epoxide	2.4	0.085	0.054	2.1	0.9:1
Oxychlordane	9	0.11	0.058	2.2	0.25:1
Trans-nonachlor	2.2	0.12	ca. 0.05	1.9	0.9:1

a Data for DDE and PCBs from U.S. FDA Diet Survey as reported by U.S. EPA (1975) and Jelinek and Corneliussen (1976); data for other chemicals derived from mean tissue levels by pharmacokinetic calculations (Hunter et al. 1969, Moriarty 1975, Nisbet 1977). Note that dietary intakes of DDE decreased six-fold between 1966 and 1973 without a correspondingly large decline in tissue storage levels (U.S. EPA 1975)

b Data from U.S. EPA Human Monitoring Survey as reported by Kutz and Strassman (1976), Kutz et al. (1976), U.S. EPA (1975).

c Data from national survey in 1975-1976 (U.S. EPA 1977b)

d Assuming a 6 kg infant ingesting 1000 ml milk containing 3.8% fat

rats, atrophy of the testes occurred. Toxic effects in humans also include weight loss, neurologic impairment, abnormal liver function, skin rash, and reproductive failure (Center for Disease Control 1976).

Kepone crosses the placenta of mice and accumulates in fetuses (Huber 1965). It has caused fetal mortality and congenital malformations in mice and rats (Chernoff and Rogers 1976). The malformations included enlarged renal pelves, undescended testes, and enlarged cerebral ventricles in rats, and clubfeet in mice. The molecular size and lipophilicity of Kepone suggest that it could cross the human placenta. Although no data on human transplacental effects are available, a potential hazard must be assumed.

Kepone is excreted in humans' (Suta 1977) and cows' (Smith and Arant 1967) milk. It accumulates in the tissues of suckling mice nursing on exposed mothers (Huber 1965). No evidence of newborn human toxicity from either mothers' milk or contaminated cows' milk is available, but a potential hazard must be assumed.

Kepone has considerable potential for chronic toxicity; it is eliminated faster than mirex (Huber 1965) but still slowly, and hence accumulates. Because of its lipophilicity, it is stored in animal tissues such as liver, brain, kidney, and fat. Chronic toxic effects on mice included weight loss, neurologic impairment, hepatomegaly, reproductive failure, reduced hematocrit, and nephropathy. Similarly, in humans it is found in liver and fat. The most frequent sign of toxicity in humans is nervousness and tremor, but opsoclonia, hepatomegaly, and abnormal sperm morphology are also prominent (Cannon et al., in press).

Kepone has mainly caused liver tumors in mice and rats, although other malignancies of the rat thyroid and reproductive system have been reported (National Cancer Institute 1976). It is therefore carcinogenic for rodents and must be considered a potential human carcinogen.

Mirex

Mirex is lethal as a single dose to rats (U.S. EPA 1978). It appears not to require metabolism in order to exert its toxicity and, in keeping with this, toxicity does not differ significantly between sexes. Thus, it is likely that it would be similarly toxic to all mammals including humans.

The subacute toxic effects most commonly observed in mammals have included weight loss, hepatomegaly, and reproductive failure (U.S. EPA 1978). An important feature of the effect on the liver is the induction of mixed

38

function oxidases (Baker et al. 1972). Again, in view of the apparent lack of a requirement for metabolism and of the documented multi-species effects, it is concluded that similar toxicity to humans would occur.

Mirex crosses the placenta of rats and accumulates in fetal tissues (Gaines and Kimbrough 1970, Khera et al. 1976). It also induces a low incidence of cataracts in transplacentally exposed fetuses (Gaines and Kimbrough 1970) as well as visceral anomalies including scoliosis, cleft palate, short tail, and heart defects (Khera et al. 1976). The molecular size and lipophilicity of mirex suggest that it could cross the human placenta. Although no data on human transplacental effects are available, a potential hazard must be assumed.

Mirex is also excreted in the milk of rats (Gaines and Kimbrough 1970), cows (Collins et al. 1974), and goats (Smrek et al., in press), and induced a high incidence of cataracts in rats nursed from exposed foster mothers (Gaines and Kimbrough 1970). Kepone, as well as other organochlorine pesticides, has been found in human milk and, therefore, it seems likely that mirex would also be excreted in milk, although none was detected at a level of sensitivity of 30 ppb in 1,436 samples taken nationwide (Suta 1977). No evidence of human newborn toxicity from either mothers' milk or contaminated cows' milk is available, but a potential hazard must be assumed.

Mirex has considerable potential for chronic toxicity since it is not metabolized, is eliminated very slowly, and hence accumulates (Ivie et al. 1974). It is stored in animal tissues such as fat, liver, and brain. Mirex is found in human fat (Suta 1977), but human chronic toxicity has not been reported.

Mirex has caused liver tumors in mice (Innes et al. 1969) and rats (Ulland et al. 1977) and therefore must be considered a potential human carcinogen.

Hexachlorocyclopentadiene

Hex is lethal as a single oral dose to mice, rats, and rabbits (Treon et al. 1955). Toxic changes occurred in brain, heart, liver, adrenal glands, kidneys, and lungs. It is also a skin irritant in rabbits and monkeys, and is sufficiently absorbed to produce toxic and lethal effects (Treon et al. 1955). The irritant effects on humans are well known (Ingle 1953). Vapors of hex are toxic to mice (Ingle 1953) and lethal to mice, rats, guinea pigs, and rabbits (Treon et al. 1955).

Repeated inhalation of hex by mice, rats, guinea pigs, and rabbits produced toxic and lethal effects (Treon et al. 1955, Industrial Bio-Test Laboratories, Inc. 1975). It is likely that it would be similarly toxic to humans.

Chronic peroral toxicity was not found in rats (Naishtein and Lisovskaya 1965), but chronic inhalation produced degenerative changes in the livers and kidneys of rats, guinea pigs, and rabbits (Treon et al. 1955). Similar toxicity to humans is likely. Hex was not found to be mutagenic to Salmonella typhimurium either with or without a rat liver microsomal activating system (Industrial Bio-Test Laboratories, Inc. 1977).

QUANTITATIVE ESTIMATES OF RISK

Kepone

The acute toxicity of Kepone is greater than that of mirex in rats. The single-dose LD_{50}s in rats, rabbits, and dogs (U.S. EPA 1978) were in the range of 71 to 250 mg/kg and, thus, did not vary more than four-fold between species. A dose of 9.6 mg/kg Kepone daily for five days to rats produces tremors. Exposure to 40 ppm Kepone in the diet for five days is sufficient to produce liver and fat deposits in mice that are 25 to 50 percent of those observed in severely intoxicated humans, (i.e., about 50 ppm in fat, [Center for Disease Control 1976]). Doubling the dose in mice results in lethality. Thus, the acutely toxic doses in animals are probably a reasonable approximation of the dose that would be acutely toxic to humans. Of present possible exposures, the one that would produce the highest single-dose levels is accidental ingestion of ant and cockroach bait. Even this would only result in ingestion of about 5 mg (Suta 1977), which for a thirty-pound child would be about 0.075 mg/kg and, thus, below the expected toxic dose.

In studies of subacute toxicity, cows suffered no ill effects at 5 ppm for 60 days (Smith and Arant 1967). Levels of 30 ppm induced tremors in mice within four weeks (Huber 1965) and 50 ppm did so in rats within three weeks (United Fruit Company 1969). Mice developed hepatomegaly on 40 ppm and were killed by 80 ppm within 32 days (Huber 1965). In a three-month study, hepatocellular changes were produced by 25 ppm in male rats and 50 ppm in females while 10 ppm caused testicular atrophy (United Fruit Company 1969). The three-month chronicity factor is 41 (Jaeger 1976). No human toxicity as a result of repeated oral intake of Kepone has been reported, but toxicity would be expected at doses comparable to those affecting other mammals. Atmospheric exposure to levels of up to 3.0 mg/m³ (Jaeger 1976) has resulted in an attack rate of 64 percent in production personnel with an average latent period of six weeks (Cannon et al., in press). Apart from exposure associated with product manufacture, the highest human atmospheric exposure would be due to the indoor use of nonenclosed insect bait, which has been estimated to produce a maximum exposure of 9 ng/m³ (Suta 1977). This is well below the nontoxic repeated doses of 15.4 µg/l in rats (United Fruit Company 1969) as well as the occupational exposures associated with toxicity.

Fetal toxicity in rats and mice was produced only at doses (10 to 12 mg/kg/day for ten days) which were also acutely toxic to the mothers (Chernoff and Rogers 1976). There is no evidence on which to determine whether subtoxic doses would affect human fetuses.

The toxicity of post-natal exposure appears not to have been studied. The stomach contents of a six-day-old mouse nursed by a mother who had been fed 40 ppm for six days contained 16.8 ppm Kepone (Huber 1965). Kepone has been found in human milk at concentrations up to 5.8 ppb (Suta 1977), well below doses that are toxic under other circumstances.

Chronic consumption of Kepone produced tissue levels up to four times the dietary level (Huber 1965) and, unlike mirex, there was a plateau in accumulation. Growth of rats and dogs was reduced at 25 ppm (U.S. EPA 1978), a level in the range of subacute toxic and carcinogenic doses. Because of the cumulative nature of Kepone, chronic toxicity at lower doses would be expected in all mammals, including humans.

Kepone produced liver tumors in rats at doses of 23 to 40 ppm and in mice at doses of 8 to 26 ppm (National Cancer Institute 1976). The mouse strain used has a significant incidence of spontaneous liver tumors and the organochlorine pesticide DDT has been demonstrated to exert a promoting (rather than initiating or carcinogenic) effect on rodent liver tumor development (Peraino et al. 1975). Thus, the relevance to human cancer risk of rodent liver tumors induced under these conditions remains to be established. Nevertheless, the possibility of some risk must be assumed. Evaluation of the level of this risk entails a quantitative extrapolation from animal experiments to human risk that is subject to a number of uncertainties (Hoel et al. 1975, Crump et al. 1976, Mantel and Schneiderman 1975, Brown 1976). The process of risk extrapolation consists basically of two steps: (1) extrapolation within the experimental animal species from the high-level test exposures to much lower levels; and (2) extrapolation of these low-exposure risks for the animal species to the risks for humans at comparable levels.

The first step requires the assumption of some mathematical model that relates the exposure level of the

41

test chemical to the observable response under consideration. Although different models can produce widely different risks (U.S. Food and Drug Administration 1971), the single-hit model, or simple linear extrapolation, should provide an upper bound to these low-exposure risks (Hoel et al. 1975). A single-hit model will therefore be used here to estimate risks from low-level exposures to both Kepone and mirex.

The occurrence of hepatocellular carcinomas or neoplastic nodules of the liver is the response to be related to the exposure level. Relating total tumor incidence to exposure would be another approach, but the inclusion of tumors unrelated to exposure could bias the estimated dose-response relationship. It should also be kept in mind that the experimental data used for these extrapolations were not designed for low-exposure extrapolations; one or two dose levels around the maximum tolerated dose (MTD) do not provide enough information on the dose-response relationship, and the published results do not include all the relevant information, such as correction factors for differential survival with treatment.

The uncertainties involved in extrapolation within the animal species may be minor compared to those in extrapolation between species. For any chemical there are likely to be species differences in absorption, metabolism, distribution, and excretion which should be taken into account in quantitative extrapolation. However, with Kepone and mirex, the similarities in disposition between species and the structurally determined resistance of Kepone and mirex to metabolism, reduce the importance of these factors. Nevertheless, extrapolating from an inbred animal strain to a heterogeneous population, such as man, presents problems. In the experimental animal, factors such as genetic disposition towards development of certain neoplasms and susceptibility to certain pharmacologic effects may be critical in determining the response. The effect on humans, as with any chemical carcinogen, can be modified by several factors other than genetic heterogeneity, such as exposure to other chemicals and diet. In the absence of comparative species information on the fate of the chemical from environmental exposure to its effects at the target site, the simplistic species-to-species extrapolation approach used here is based on an assumed equivalence of total exposure per unit of body weight (National Research Council 1975). Since it is not known which animal species is most comparable to man, extrapolation is done for every set of experimental data in order to provide a range of estimated human risks.

Because of the incompleteness of the experimental data, the use of a single-hit model, and the uncertainties involved in this extrapolation process, these estimates of human risk should not be considered as best estimates but rather as crude upper confidence bounds to the true risk. This assumes that the true dose-response curve is convex at low dose levels and that man is no more sensitive than the experimental animals.

The extrapolations based on the results of the National Cancer Institute mouse bioassay were performed for each sex separately since male mice appeared to be more sensitive than females. The mice were exposed to Kepone for 80 weeks at two dose levels. The dietary levels changed over time, but the average exposure levels were 20 and 23 ppm for males and 20 and 40 ppm for females. The total exposures per unit of body weight are approximately:

Males: $20 \times 10^{-6} \times (4 \times 10^3)$ mg (diet)/day x (7 x 80) days $\div 0.03$ kg (body weight) = 1,490 mg/kg

$23 \times 10^{-6} \times (4 \times 10^3)$ mg/day x (7 x 80) days $\div 0.03$ kg = 1,720 mg/kg

Females: $20 \times 10^{-6} \times (4 \times 10^3)$ mg/day x (7 x 80) days $\div 0.03$ kg = 1,490 mg/kg

$40 \times 10^{-6} \times (4 \times 10^3)$ mg/day x (7 x 80) days $\div 0.03$ kg = 2,980 mg/kg

When the single-hit model is fit to the experimental data, the relationship between total exposure and incidence of hepatocellular carcinoma attributable to low-level exposure to Kepone is approximately proportional to exposure, that is $R = a \times E$, where R is the lifetime attributable risk at a total exposure of E milligrams of Kepone per unit of body weight, and a is the proportionality constant estimated from the experimental animal data. For males and females, these proportionality constants are estimated as 9.3×10^{-4} and 2.9×10^{-4}, respectively. Because of statistical variation in the experimental data, the upper 95-percent confidence limits on these constants are 1.2×10^{-3} and 3.9×10^{-4}. Besides the uncertainties inherent in any animal bioassay, the high response rates (over 80 percent for the treated male mice) add to the difficulty of extrapolation. Since it is difficult to determine whether or not lower dose levels would have resulted in similar response rates, the risk estimates based on these results should be given somewhat less weight than the others.

The estimated human risk for a calculated cohort of 10 million persons exposed to 1 μg/day (see section on Distribution of Exposures) is given in Table 2.4. This extrapolation assumes that present exposures will hold constant even though this may not in fact be true. The

TABLE 2.4 Estimated Number of Lifetime Cancer Cases in a Cohort of 10 Million People Exposed to 1 µg/day of Kepone or Mirex throughout their Lifetime

Chemical	Source of Experimental Results	Animal Species/Sex	Estimated Number of Cancer Cases*
Kepone	National Cancer Institute (1976)	Male Mice	3,400 (4,400)
		Female Mice	1,100 (1,400)
		Male/Female Rats	510 (950)
Mirex	Innes et al. (1969)	Male Mice	840 (1,500)
		Female Mice	1,800 (2,400)
	Ulland et al. (1977)	Male/Female Rats	330 (470)

*The numbers in parethenses are 95% statistical upper confidence limits on the estimated numbers of cancer cases.

44

daily exposure rate was converted to a total lifetime exposure and, assuming an average 70-kg human whose average lifespan is 70 years, then a person's total lifetime exposure per unit of body weight is 0.365 mg/kg. Table 2.4 gives the number of these 10 million people estimated to develop at some time in their lifetime cancer that is attributable to this chronic exposure. The numbers are calculated from the relationship $R = a \times E$ where a is the proportionality constant estimated from the experimental animal data and E is the assumed human exposure level. As an example, on the basis of results for male mice exposed to Kepone, the estimated human lifetime risk attributable to a constant daily exposure of 1 µg would be $R = 9.3 \times 10^{-4} \times 0.365 = 3.4 \times 10^{-4}$, which implies an expected 3,400 cases from a cohort of 10 million exposed people. The linear nature of this extrapolation relationship means that since risk is proportional to dose, calculations would also apply to a cohort in which the varying exposure among its members averages at 1 µg/day. Thus, Table 2.4 can be used to provide the estimated numbers of cancer cases from other exposure situations. For example, the numbers in the table also represent estimated cancers from a daily exposure of 10 µg to a population of one million persons, while for an exposure of 10 µg/day to 100,000 persons, the estimated number of cancers would be 10 percent of those given in the table. Other exposure levels to other populations can also be easily derived. It should be reemphasized that these estimates have been derived under the assumptions of no threshhold and a single-hit dose-response relationship. Since the single-hit model is approximately linear at low-dose levels and linearity is an upper bound for a true convex dose response, these assumptions provide "worst case" estimates of the low-dose risk. Therefore, assuming man is no more sensitive than the experimental animals, these risk estimates should be considered as upper confidence bounds to the true low-dose risks, whch may even be zero.

A similar risk extrapolation for the possible carcinogenic effects on humans of allowing the estimated Kepone stocks of 538 lb to be used over a three year period was completed in 1976 by the EPA Carcinogen Assessment Group (U.S. EPA 1976a). In their view, the major concern was with accidental ingestion of trap contents or paste by children. In the "worst case" in which a child would eat six 3 oz tubes of ant paste over three years, the lifetime risk with such exposure was a one in 400 chance of developing cancer. A second analysis, in which data from the EPA Pesticide Episode Reporting System was used to predict that 3,750 children might each eat about half an ant trap, yielded a calculation of a probability of one-twelfth of one case of cancer. In a third analysis, an attempt was made to estimate parametrically the effect of all routes of exposure. Using a linear no-threshold extrapolation, the resulting number of cancer cases was calculated to be 540 multiplied by the proportion of Kepone that reaches humans. No estimate of this proportion was available, but as an example, if one-tenth of one percent of the total stock ultimately reached humans, then about one-half a case of cancer would be projected to result.

The quantitative extrapolation methods used here differ somewhat from those employed in the risk assessment performed by the EPA Carcinogen Assessment Group. Their analysis was based only on the experimental results from the high-dose male mice, and was extrapolated to humans on an equivalent dietary concentration rule (Mantel and Schneiderman 1975). We feel that a better estimate of the proportionality constant can be obtained from all the experimental results, not just those of the treated group (which gives the highest value for this constant), and that different exposure durations can best be incorporated into the extrapolation by the rule of equivalent total dose per unit of body weight. Since man's exposed lifetime is greater than the animal's exposure duration, our estimated risks are approximately six times greater than those obtained by EPA. In addition, we have attempted to provide an indication of the variability introduced into the extrapolation process by the sex and strain of the experimental animals, and by statistical variation in the experimental results. Furthermore, as noted above, our analysis is based upon constant exposure levels whereas the Carcinogen Assessment Group used limited exposures.

The extrapolations from the data on exposure of rats to Kepone (National Cancer Institute 1976) are based on males and females combined, since there was little evidence of a difference in sensitivities. The rats were exposed to Kepone in their diet for 80 weeks at two dose levels. The dietary levels changed over time, but the average exposure levels were 8 and 24 ppm for males and 18 and 26 ppm for females. Assuming an average 400-g rat eating 20 g/day, the total exposures per unit of body weight are approximately:

Males	Females
220 mg/kg	500 mg/kg
570 mg/kg	730 mg/kg

When the single-hit model is fit to these data, the proportionality constant for both males and females is 1.4×10^{-4} with a 95-percent statistical upper confidence limit of 2.6×10^{-4}. The number of cancers estimated to occur in the assumed cohort of 10 million persons exposed to 1 µg/day is given in Table 2.4.

Mirex

The acute toxicity of mirex has been quantified only in
rats. The lowest oral single dose which will kill a rat is
400 mg/kg for males and 500 mg/kg for females (Gaines and
Kimbrough 1969). The single-dose LD_{50} in Sherman strain
rats was 740 mg/kg for males and 600 mg/kg or 365 mg/kg for
females (U.S. EPA 1978). This LD_{50} is higher than that of
Kepone or DDT (Gaines and Kimbrough 1969) indicating a
lesser acute toxicity than for those agents. Single doses
of 1 mg/kg were well tolerated by rhesus monkeys (Wiener et
al. 1976) and led to fat levels of 1.7 to 5.8 ppm. A
comparison of these levels with documented levels in humans
in the general population of up to 1.3 ppm in adipose tissue
of individuals allows the assumption that similar acute
doses could be tolerated by humans, but that toxicity would
occur at the doses that are toxic to other mammals.

Subacute effects have been observed at rather low levels
of exposure. In rats, 1.0 ppm in the diet caused induction
of cytochrome P-450 within 14 days (Baker et al. 1972).
Toxicity was not observed in rats at dietary levels of 50
ppm (3.1 mg/kg/day) (Gaines and Kimbrough 1969) whereas mice
were killed within 14 days by feeding at that level (Ware
and Good 1967). Oral administration to pregnant rats on
days 6 to 15 of gestation produced toxic effects with daily
doses of 3.0 to 12.5 mg/kg (Khera et al. 1976). These doses
resulted in adipose levels of 48 to 281 ppm. The 90-dose
LD_{50} in rats is 6.0 mg/kg, giving a chronicity factor (ratio
of the single dose LD_{50} to the 90-dose LD_{50} [Hayes 1967] of
60.8 [Gaines and Kimbrough 1970]). This was very high in
comparison to chronicity factors of 5.4 for DDT and 12.8 for
dieldrin, indicating a highly cumulative effect. Thus, the
current U.S. Food and Drug Administration food tolerances of
0.3 ppm in fish and shellfish and 0.4 ppm in crabs are close
to dietary levels producing biochemical abnormalities in
other mammals, but below subacute toxic levels.
Furthermore, it is unlikely that humans would consume a diet
uniformly contaminated to that level. In fact, the highest
human adipose level detected in the EPA monitoring program
(i.e., 1.3 ppm) was 1/160 of that (i.e., 211 ppm) produced
in rats by feeding 3 ppm for six months (Ivie et al. 1974).
Therefore, the levels which have been permitted in human
food are unlikely to cause subacute toxicity, but it would
be expected to occur at doses similar to those active in
other mammals.

Transplacental exposure of rat fetuses carried by
mothers fed 5 ppm resulted in a 1.6-percent incidence of
cataracts (Gaines and Kimbrough 1970). Treatment of
pregnant rats on days 6 to 15 of gestation resulted in
visceral anomalies in the fetuses with doses of 6.0 and 12.5
mg/kg, but not with 1.5 and 3.0 mg/kg. Although the doses
producing malformations were also toxic to the mothers, 5

47

ppm was well tolerated and, thus, a teratogenic hazard to humans at nontoxic doses must be presumed.

Post-natal exposure of rats as a result of nursing on foster mothers fed 5 ppm for 73 days before nursing produced a 38-percent incidence of cataracts in the newborns (Gaines and Kimbrough 1970). The mirex content of milk recovered from the stomachs of the newborns was 11.3 ppm. Mirex has not been found in human milk (Suta 1977), and the concentration in whole milk from cows grazing in a treated pasture was 0.002 to 0.007 ppm (U.S. EPA 1978) or less than 0.3 ppb within six months after treatment (Hawthorne et al. 1974). Thus, any potential human exposure from this source is well below the toxic levels in animal studies.

Chronic dietary consumption of mirex results in tissue accumulations of up to 120 times the intake levels (Ivie et al. 1974). Furthermore, unlike other organochlorine pesticides (e.g., DDT, dieldrin) there seems to be no plateau in tissue residue concentration (Ivie et al. 1974). However, doses of up to 30 ppm to rats for 16 months (Ivie et al. 1974) and 1 mg/kg to goats for 61 weeks (Smrek et al., in press) were reported to be without toxicity. Human fat concentrations of up to 1.3 ppm (U.S. EPA 1978) indicate very low levels of exposure that would be below toxic doses in animal studies.

Mirex is hepatocarcinogenic in mice (Innes et al. 1969) and rats (Ulland et al. 1977). The mouse strains used have a significant incidence of spontaneous liver tumors and the dose in rats was sufficient to induce hepatotoxic changes. With the same reservations described in the extrapolation of human risk from the Kepone data and following the same procedures, extrapolations were performed on the results obtained in these studies.

In the testing of mirex in mice (Innes et al. 1969), the two strains of mice used were equally sensitive and therefore were combined in the analysis. However, since females were apparently more sensitive to mirex than males, the extrapolations are based on each sex separately. The mice were exposed when they were between 7 and 28 days old to mirex at a level of 10 mg/kg by stomach tube, and then at 26 ppm in their diet throughout the remainder of the study (terminal necropsy at age 59 weeks for males and 69 to 70 weeks for females). For purposes of computing total exposure per unit of body weight, it is assumed that a young mouse weighs 10 g and an adult mouse weighs 30 g while consuming 4 g of food daily. The total exposure for males is:

$$10 \text{ mg/kg/day} \times 0.01 \text{ kg} \times 21 \text{ days} = 2.1 \text{ mg} +$$
$$26 \times 10^{-6} \times (4 \times 10^3) \text{mg/day} \times (7 \times 55) \text{ days} = \underline{40.0 \text{ mg}}$$
$$= 42.1 \text{ mg}$$

The total exposure for females is 49.8 mg because of their longer exposure at 26 ppm in the diet. Those exposures per unit of body weight are approximagely 1,400 mg/kg and 1,660 mg/kg, respectively.

For males and females the proportionality constants are estimated as 2.3×10^{-4} and 5.0×10^{-4}, respectively. Because of statistical variation in the experimental data, the upper 95-percent confidence limits on these constants are 4.0×10^{-4} and 6.5×10^{-4}, respectively. The estimated human risk for the assumed cohort of 10 million persons constantly exposed to 1 µg/day (see section on Distribution of Exposures) is given in Table 2.4.

The extrapolations from the data on exposure of rats to mirex (Ulland et al. 1977) are based on both males and females since there was no evidence of different sensitivities. The animals were exposed at 40 and 80 ppm in their diet for 9 weeks and at 50 and 100 ppm from the 10th week through 18 months. Assuming an average 400-g rat eating 20 g/day, the total exposures are:

$$40 \times 10^{-6} \times (20 \times 10^3) \text{ mg/day} \times (7 \times 9) \text{ days} = 50.4 \text{ mg} +$$
$$50 \times 10^{-6} \times (20 \times 10^3) \text{ mg/day} \times (7 \times 69) \text{ days} = \underline{483.0 \text{ mg}}$$
$$533.4 \text{ mg}$$

for the lower dose animals and 1,066.8 mg for the higher dose group. These exposures per unit of body weight are approximately 1,330 mg/kg and 2,660 mg/kg. When the single-hit model is fit to these data, the proportionality constant for both males and females is estimated as 9.0×10^{-5} with an upper 95-percent confidence limit of 1.3×10^{-4}. The estimated human risk for the assumed cohort of 10 million persons exposed to 1 µg/day is given in Table 2.4.

Hexachlorocyclopentadiene

The oral LD_{50} of a single dose of hex was 430 mg/kg in mice (Lu et al. 1975), and 584 mg/kg in rats (Velsicol Chemical Corporation 1976). Estimation of acute toxicity to rats yielded LD_{50}s for a single oral dose as 505 mg/kg (Treon et al. 1955), and 600 mg/kg (Naishtein and Lisovskaya 1965). The minimum lethal dose in female rabbits is 420 mg/kg (Treon et al. 1955). The similarity in these toxic doses and the absence of known metabolism suggest that hex would be similarly toxic to humans.

Exposure of rats to concentrations of 0.0006 and 0.014 mg/l of air for six hours per day, five days per week for four weeks resulted in toxic liver changes, but administration of 30 to 300 ppm in the diet produced no subacute toxicity in a 90 day study (Industrial Bio-test Laboratories, Inc. 1975).

No chronic toxicity was noted in rats administered 0.0002 and 0.00002 mg/kg for up to six months (Naishtein and Lisovskaya 1965). Rats, guinea pigs, and rabbits survived inhalation of 0.15 ppm (0.0017 mg/l) for seven hours per day five days per week over a 216 day interval, but developed pathologic changes in the livers and kidneys (Treon et al. 1955).

Owing to a lack of information, no extrapolation was made concerning the carcinogenic risk of human exposure to hex.

CHAPTER 3

ENVIRONMENTAL CONSIDERATIONS

CRITERIA FOR ASSESSMENT

Evaluation of the significance of environmental contamination should be based upon criteria that allow objective assessment of the characteristics of a given pollutant. Such criteria should be developed according to known ecological principles and proven environmental distribution of the contaminant in question. It should be acknowledged that relatively few natural systems have been described in adequate detail to allow the development of predictive models of impact. The following is a brief review of suggested criteria for assessing the environmental impact of Kepone, mirex, and hex.

Field and laboratory studies have established certain characteristics of Kepone, mirex, and hex, and have indicated that all three compounds can be bioconcentrated and biomagnified (U.S. EPA 1978). The compounds tend to show considerable environmental persistence (although our knowledge of hex remains too incomplete for a definitive finding on this point). Hex can be highly toxic to various organisms, although little is known about its chronic effects and environmental cycling (Lu et al. 1975). The toxicity of Kepone and mirex to various species, together with their high level of environmental stability, present a potential threat at the system level. Since various areas have been contaminated with Kepone and mirex, and both compounds have a proven potential for transboundary movement and biological activity in various organisms, any assessment of impact should include an evaluation at the population and community levels. This would include analysis of population dynamics of exposed species (Hurlbert et al. 1972), interpopulation response (Mosser et al. 1972, Mosser et al. 1974), potential alteration of energy transfer mechanisms (Hurlbert et al. 1972), and community reaction in affected areas (Livingston et al., in press).

The complexity of natural systems (e.g., short- and long-term variation, multiple energy transfer pathways, exposure to different stressors or forcing functions), together with a relatively incomplete knowledge of the fate

51

of persistent organochlorine compounds in the environment, have led to a poor understanding of pesticide impact at the system level (Livingston 1977). Variation (species-specific) of toxic effects on exposed populations, secondary (chronic) effects due to bioconcentration and biomagnification, interruption of predator-prey relationships and competitive interactions, changes in processes of selection, and eventual alteration of genomes of individual populations as a result of direct or indirect alterations of the physicochemical habitat through time are all involved in the potential for environmental impact. Because of the possibility of changes in pathways of energy flow via differential impact on sets of interacting populations, a system can be considerably altered without the change being detected by current methods of environmental analysis. Lacking a comprehensive knowledge of a particular system over sufficiently long periods of time, such questions are rarely addressed and a vast majority of resulting "predictive" models become mere speculation. Many aquatic systems are characterized by a highly complex series of homeostatic mechanisms that continuously reflect time-labile combinations of natural forcing functions and stressors. Often, a given set of conditions will result in a reduction of species richness and/or associated changes in species composition, and this change is transmitted to trophic interactions; consequently, a loss of biological information at any given level can become critical. This can lead to disruption of the system and loss of useful productivity. Variable sensitivity from one ecosystem to the next complicates the process of evaluation. When toxic agents which are environmentally persistent have the potential for adverse effects at various levels of biological organization, it becomes difficult to determine "threshold" levels (often derived from extrapolation of acute and toxic bioassay) of impact in quantitative terms. It follows that compounds such as Kepone, mirex, and hex should be evaluated according to the above criteria if environmental impact is to be assessed.

ACUTE AND CHRONIC EFFECTS

Terrestrial Organisms

Kepone

The lack of comprehensive information concerning the toxicology of Kepone precludes serious generalization. Various subacute effects, including a spectrum of neurological and reproductive disorders, have been noted in experiments with birds (U.S. EPA 1978). Chronic effects of Kepone on mammals are also evident and include various (multiple) levels of activity. There are indications that both Kepone and mirex cause complete inhibition of LDH at

52

concentrations of 10 ppm (in vitro determinations, Hendrickson and Bowden 1975). Other effects of Kepone on enzyme systems that have been documented at levels comparable to actual bioaccumulation in the field imply potential adverse impact of Kepone on energy production and muscle function (U.S. EPA 1978). Lipophilic chemicals such as Kepone and mirex may interfere with various cellular processes including functions of DNA and the endoplasmic reticulum. Chronic tests with mammals at Kepone levels ranging from 5 to 80 ppm have resulted in various effects including tremors, weight depression, and chronic reproductive incapacity (U.S. EPA 1978). Overall, Kepone appears from the available literature to be capable of causing adverse chronic effects in various terrestrial organisms at relatively low levels of exposure. There is a need for analysis of effects on terrestrial assemblages exposed to Kepone at indeterminate though presumably continuous levels (i.e., Puerto Rico).

Mirex

Mirex residues in terrestrial assemblages appear to be dependent on trophic relationships (Collins et al. 1974). Organisms such as crickets, oil-loving ants, and ground beetles are directly affected, and predaceous arthropods such as spiders retain relatively high mirex residues (U.S. EPA 1978). Based on mortality (mirex is considered moderately toxic), this organochlorine does not appear to be a threat to many nontarget species at actual application rates, 4.2 g/ha (1.7 g/acre), although elimination of susceptible populations found in some work at the population and community levels (Reagan et al. 1972) indicates potential indirect or chronic effects (as discussed below). Less is known about the accumulation of mirex in vertebrates, although there are indications that it is not concentrated by amphibians and reptiles as readily as by birds and mammals (U.S. EPA 1978). Birds are not extremely sensitive to acute toxic effects of mirex; however, the relatively high levels of residues in wild birds in treated areas and the lack of data about the possibility of reproductive effects of this pesticide on natural populations remains a potential problem for continued broadcast use of mirex in the program for control of the imported fire ant. This is related to the proven stability and persistence of this compound in the environment (U.S. EPA 1978). While acute mirex toxicity to mammals is relatively low, chronic exposure produces a wide range of effects; the chronicity factor (Hayes 1967) is one of the highest observed for any pesticide, largely because of the highly cumulative nature of mirex in biological systems and the fact that it is metabolized or excreted at extremely slow rates (U.S. EPA 1978). The pervasiveness of chronic physiological and biochemical disorders induced by mirex in

53

experiments with various vertebrate species, together with
its capacity for bioaccumulation, would tend to bring into
question the widespread (broadcast) use of this pesticide.
The accumulation of mirex in humans near areas of fire ant
control (Suta 1977) substantiates the need for further work
on chronic effects of mirex on various vertebrate species.

Hexachlorocyclopentadiene

There are too few data about the impact of hex on
terrestrial organisms to draw any conclusions. In addition
to its germicidal and fungicical properties (Cole 1954), hex
is a herbicide (small grasses and weeds). While the
environmental fate of hex in terrestrial systems appears
similar to that of certain other organochlorines (Lu et al.
1975), specific effects remain unclear. Since hex can enter
the terrestrial environment as a dust or volatile gas (thus
affecting organisms by cutaneous contact, through
respiration, and/or drinking), the potential for entry into
biological systems is high (U.S. EPA 1978). However, it
remains difficult to assess acute or chronic impact of this
compound at the system level, although the potential for
environmental persistence and toxic effects remains high.

Aquatic Organisms

Kepone

Kepone causes adverse effects on algal growth at low
concentrations with 24-hour EC_{50}s of four species of marine
unicellular algae ranging from 0.35 to 1.0 ppm (Walsh et al.
1977). In addition to being highly toxic to various aquatic
invertebrates, Kepone causes chronic effects such as reduced
shell growth, loss of equilibrium, and impaired reproductive
capability (U.S. EPA 1978). Bioconcentration, probably
through feeding or exposure to contaminated sediments, can
be high under natural conditions (e.g., average
concentrations of 0.3 to 3.0 ppm in blue crabs, 0.04 to 2.0
ppm in finfishes [Suta 1977]). Chronic tests indicate a
life-cycle (19 day) LC_{50} of 1.4 $\mu g/l$ in mysids with Kepone-
induced growth inhibition in mysids at 0.07 $\mu g/l$ (Hansen et
al. 1977b). Again, species-specific variability of chronic
impact has been demonstrated although the available
information indicates that Kepone is biologically active
and, in many cases, causes a range of adverse chronic
effects in disparate invertebrate species at low levels of
exposure. Experiments with fishes indicate that although
Kepone has low acute toxicity to some estuarine species, to
others it can be highly toxic at relatively low exposure
concentrations (low ppb) (U.S. EPA 1978). There are
indications that the long-term effects of Kepone on juvenile
fishes can be extensive at concentrations of less than 1 ppb

(U.S. EPA 1978). Mechanisms of physiological effects remain
poorly understood although ATPase inhibition has been
established (U.S. EPA 1978).

There have been a number of recent contributions to the
Kepone literature. Walsh et al. (1977) found that various
unicellular marine algae are capable of concentrating Kepone
from water. Although such organisms probably are not killed
at the concentrations observed in the field, they could
serve as agents of transfer and accumulation at higher
trophic levels. Schimmel and Wilson (1977) found that
Kepone was acutely toxic to shrimp and fishes but not to
crabs, with bioconcentration occurring in all organisms
tested. Hansen et al. (1977a) found that acute tests
greatly underestimated the hazard of Kepone to fish and that
concentrations as low as 0.08 µg/l affected certain
processes such as growth. Life-history studies showed that
there was abnormal development of embryos taken from adult
fish exposed to 1.9 µg/l. In such studies, concentration
factors up to 20,000 times exposure concentrations were
demonstrated. Couch et al. (1977) found that Kepone can
induce scoliosis in fish as part of a poisoning syndrome
involving histological alteration. There were indications
that the nervous system is a primary target for this
pesticide. Severity of effect appears to depend on duration
of exposure. Various studies indicate that chronic toxicity
and the bioconcentration potential of Kepone are factors
that will have to be taken into consideration in any
assessment of impact due to environmental contamination. At
present Kepone concentrations of chronic (laboratory) effect
are well within known levels of Kepone found in runoff from
banana fields in Puerto Rico, or residue concentrations in
the James River system. However, system level impact
remains undetermined due to a lack of background
information.

Mirex

The biological significance of mirex is related to its
chemical characteristics. Modes of transfer into living
systems (described above) are important to an understanding
of the impact of this insecticide on aquatic organisms.
Mirex reduces productivity of green algae; various species
of phytoplankton can concentrate the pesticide and thus may
serve as passive agents of transfer to consumer organisms
(U.S. EPA 1978). Although few adverse effects on algal
species have been demonstrated, this area of research needs
further analysis especially with respect to the ecological
implications of some recent findings. Mirex does not appear
to have pronounced acute effects on fishes in a range of
concentrations found in treated areas. However, dose-
dependent secondary effects such as bacterial infection
(goldfish), and growth inhibition (bluegills, catfish),

55

appear to be related to mirex accumulation (U.S. EPA 1978).
Life-history studies and results concerning synergistic
interactions (Koenig 1977) indicate that further work along
this line would be instructive.

Various forms of freshwater and estuarine arthropods are
extremely sensitive to mirex, with high mortality at
concentrations as low as 0.1 ppb (U.S. EPA 1978). Juvenile
forms are often most susceptible and larval stages of some
species show adverse sublethal reactions at concentrations
as low as 0.01 ppb. Irritability and mortality have often
occurred after exposure; this is the so-called delayed
effect which is a distinctive characteristic of mirex in a
variety of aquatic species. Although certain factors (age,
size, species, physicochemical factors, etc.) influence the
form and degree of response (including irritability, loss of
equilibrium, paralysis, and death), mirex evidently is an
effective biocide for various forms of aquatic invertebrates
(U.S. EPA 1978). This should be an important consideration
in any evaluation of the environmental impact of mirex.
Once again, however, the environmental significance of mirex
in aquatic systems remains undetermined, although the
potential for adverse effects at the population and
community levels remains high.

Hexachlorocyclopentadiene

There is little information concerning the effects of
hex on aquatic organisms. Bioassays indicate that levels
approximating 1 ppm are acutely toxic to fish although
little is known of chronic effects. The usual species-
specific variability in acute toxicity is evident although
most of the fish LC_{50}s range in the low ppm (U.S. EPA 1978).
Although the potential of hex for persistence and
bioconcentration in aquatic systems appears good (based on
its general chemical characteristics), the lack of
documentation concerning acute and chronic toxicity
precludes a definitive statement regarding environmental
implications. The relatively uneven toxicity data, with no
real attention to aquatic plants and invertebrates, indicate
that hex can be highly toxic (a 1956 U.S. Department of
Health, Education, and Welfare report suggests "safe"
concentrations of 5 to 10 ppb). There is a clear need for
more information concerning the effects of hex on aquatic
organisms. This is especially true when the past history of
the organochlorine pesticides is taken into consideration.

ECOLOGICAL IMPLICATIONS

An important consideration in any impact determination
involves the range of ecological changes which occur as
defined within the limits of natural spatial and temporal

56

variability. Often, background variation disallows clear-
cut evaluation of the environmental significance of a given
pollutant. While it is true that chronic effects at the
population and community levels may be more profound than
the more readily apparent acute effects (usually denoted by
mortality), extrapolation of experimental data to field
situations can be misleading. There are complications due
to synergism and temporal variation of impact which
complicate direct delineation of causal relationships.
Various ecological approaches for such estimates are
available, including direct environmental sampling programs
(lengthy and costly), manipulation of simulated artificial
laboratory communities, the use of indicator species or
species associations, experimental (field) approaches
including caging, predator elimination, etc., or a
combination of such techniques. Such studies are needed to
address the criteria listed at the beginning of this
chapter.

Any evaluation of pesticide use should address several
primary functions: environmental persistence; potential for
bioaccumulation and biomagnification; known levels and
distribution of residues; and experimentally established
effects (acute and chronic) for a range of representative
species. The variability of species susceptibility, a basic
lack of information concerning pesticide fate, and multiple
factor interaction all tend to complicate specific
interpretations of impact. The broadcast use of
conservative pollutants can lead to the development of
resistant species in addition to potential adverse effects
on susceptible or endangered species.

There are indications that the use of mirex can cause
changes at the community level. Although most of the
studies associated with the widespread use of mirex have
offered largely anecdotal and qualitative evidence
concerning population and community reactions. However, the
U.S. Department of Agriculture has determined that routine
applications of mirex can kill various nontarget species
including oil-loving ants, spiders, beetles, and crickets
(unpublished USDA report 1976). According to S.D. Hensley
(U.S. Department of Agriculture and Louisiana State
University, personal communication 1977), application of
mirex or heptachlor for elimination of fire ants has
resulted in nontarget pest population resurgence and/or
escalation of minor pests to major pest status in several
Louisiana agroecosystems. Reagan et al. (1972) documented
increased infestation of the sugarcane borer (_Diatraer
saccharalis_) due to mirex-induced reductions of various
arthropod predators which normally feed on such pests.
Recent studies show that uptake and accumulation of mirex
can cause reductions in seed germination, seedling
emergence, and growth in several plant species (Rajanna and
de la Cruz 1975). This would indicate more pervasive

effects than toxicity studies or residue surveys would show, although the data remain thin. There is even less information concerning potential effects on natural aquatic systems. Lowe et al. (1971) point out, however, that mass mortality of estuarine organisms is unlikely due to the (delayed) mechanism of mirex effect. Evidently, the ecological implications of the widespread use of mirex remain relatively unknown although there are indications that effects on nontarget organisms could cause systemic changes not elucidated by previous studies.

The above should be viewed within the context of the overall impact of the imported fire ant. This includes its status as an agricultural pest and as a factor in human health problems. Since the ecological role of the fire ant represents a mixture of advantageous and detrimental functions with total eradication a remote possibility, widespread control through broadcast use of mirex remains tenuous. From an environmental point of view, such use of this pesticide remains dangerous and unpredictable. The need for control should thus stimulate a continued search for viable alternatives which do not aggravate the very problems that are to be corrected. As yet, this has not been subjected to a scientific perspective of total environmental impact but the proven stability and toxicity of this compound would necessitate such an approach if responsible action is to be taken.

When analyzed according to the above criteria, there are simply too few data for cogent generalizations concerning the environmental impact of Kepone, mirex, and hex. In relative terms, there is a growing though incomplete literature concerning Kepone; there is a substantial body of knowledge of the acute and chronic effects and residue distribution of mirex; and little is known with regard to the environmental significance of hex.

REFERENCES

Bahner, L.H. and D.R. Nimmo (1977) Uptake of Kepone from
    sediments by estuarine organisms. Paper presented at
    Fourth Biennial International Estuarine Research
    Conference, Mt. Pocono, Pennsylvania, 1977.

Baker, R.C., L.B. Coons, R.B. Mailman, and E. Hodgson (1972)
    Induction of hepatic mixed function oxidases by the
    insecticide mirex. Environmental Research 5:418-424.

Bender, M.E., R.J. Huggett, and W.J. Hargis (1977) Kepone
    residues in Chesapeake Bay biota. Paper presented at
    Fourth Biennial International Estuarine Research
    Conference, Mt. Pocono, Pennsylvania, 1977.

Blanchard, J. (1976) Kepone fact sheet. (Internal EPA
    memorandum, September 13, 1976)

Brooks, G.T. (1974) Chlorinated Insecticides, Volumes I and
    II. Cleveland, Ohio: CRC Press, Inc.

Brown, C. (1976) Mathematical aspects of dose-response
    studies in carcinogenesis--the concept of threshold.
    Oncology 33(2):62-65.

Cannon, S.B., J.M. Veazey, R.S. Jackson, V.W. Burse, C.
    Hayes, W.E. Straub, P.J. Landrigan, and J.A. Liddle (In
    press) Epidemic Kepone poisoning in chemical workers.
    American Journal of Epidemiology.

Carlson, D.A., K.D. Konyha, W.B. Wheeler, G.P. Marshall, and
    R.G. Zaylskie (1976) Mirex in the environment: Its
    degradation to Kepone and related compounds. Science
    194(4268):939-941.

Center for Disease Control (1976) Kepone Poisoning -
    Virginia. (Internal memorandum EPI-76-7-3, October 18,
    1976)

Chernoff, N. and E.H. Rogers (1976) Fetal toxicity of Kepone
    in rats and mice. Toxicology and Applied Pharmacology
    38:189-194.

Cole, E.J. (1954) Treatment of sewage with hexachlorocyclopentadiene. Applied Microbiology 2:198-199.

Collins, H.L., G.P. Markin, and J. Davis (1974) Residue accumulation in selected vertebrates following a single aerial application of mirex bait (Louisiana, 1971-72). Pesticides Monitoring Journal 8:125-130.

Couch, J.A., J.T. Winstead, and L.R. Goodman (1977) Kepone-induced scoliosis and its histological consequences in fish. Science 197:585-587.

Cripe, C.R. and R.J. Livingston (1977) Dynamics of mirex and its principal photoproducts in a simulated marsh system. Archives of Environmental Contamination and Toxicology 5:295-303.

Crump, K.S., D.G. Hoel, C.H. Langley, and R. Peto (1976) Fundamental carcinogenic processes and their implications for low dose risk assessment. Cancer Research 36:2973-2979.

Edwards, C.A. (1966) Persistent pesticides in the environment. Pages 1-67, Critical Reviews of Environmental Control 1. Cleveland, Ohio: CRC Press, Inc.

Equitable Environmental Health, Inc. (1976) Literature Review of the Ecological Effects of Exposure to C-56 (Hexachlorocyclopentadiene). Prepared for Hooker Chemicals and Plastics Corporation, July, 1976. Woodbury, N.Y.: Equitable Environmental Health, Inc.

Ferguson, W.S. (1975) Letter to S.R. Wassersug, Enforcement Division, U.S. Environmental Protection Agency, Region III, Philadelphia, Pennsylvania. September 12, 1975.

Frank, R., K. Montgomery, H.E. Braun, A.H. Berst, and K. Loftus (1974) DDT and dieldrin in watersheds draining the tobacco belt of southern Ontario. Pesticides Monitoring Journal 8:184-201.

Gaines, T.B. and R.D. Kimbrough (1969) The oral toxicity of mirex in adult and suckling rats. Toxicology and Applied Pharmacology 14:631.

Gaines, T.B. and R.D. Kimbrough (1970) Oral toxicity of mirex in adult and suckling rats. Archives of Environmental Health 21:7-14.

Garnas, R.L., A.W. Bourquin, and P.H. Pritchard (1977) The fate and degradation of 14C-Kepone in estuarine microcosms. Environmental Research Laboratory. Gulf

Breeze, Fla.: U.S. Environmental Protection Agency. (Unpublished Report No. 351)

Gilman, A.P., G.A. Fox, D.B. Peakall, S.M. Teeple, T.R. Carroll, and G.T. Haymes (1977) Reproductive parameters and egg contaminant levels of Great Lakes herring gulls. Journal of Wildlife Management 41(3):458-468.

Guyer, G.E., P.L. Adkisson, K. DuBois, C. Menzie, H.P. Nicholson, and G. Zweig (1971) Toxaphene Status Report. Special report to EPA Hazardous Materials Advisory Committee, November 1971. Washington, D.C.: U.S. Environmental Protection Agency.

Hansen, D.J., L.R. Goodman, and A.J. Wilson, Jr. (1977a) Kepone: Chronic effects on embryo, fry, juvenile, and adult sheepshead minnows (Cyprinodon variegatus). Chesapeake Science 18:227-232.

Hansen, D.J., D.R. Nimmo, S.C. Schimmel, G.E. Walsh, and A.J. Wilson, Jr. (1977b) Effects of Kepone on estuarine organisms. Environmental Research Laboratory. Gulf Breeze, Fla.: U.S. Environmental Protection Agency. (Unpublished paper, Contract No. 311)

Hawthorne, J.C., J.H. Ford, C.D. Loftis, and G.P. Markin (1974) Mirex in milk from southeastern United States. Bulletin of Environmental Contamination and Toxicology 11(3):238-240.

Hayes, W.J., Jr. (1967) The 90-dose LD_{50} and a chronicity factor as measures of toxicity. Toxicology and Applied Pharmacology 11:327-335.

Hayes, W.J., Jr. (1975) Toxicology of Pesticides. Baltimore: Williams and Wilkins.

Hendrickson, C.M. and J.A. Bowden (1975) A proposed mechanism for the in vitro inhibition of NADH-linked dehydrogenases by halogenated hydrocarbon pesticides: Evidence for an "association complex" for lactic acid dehydrogenase. Federation Proceedings. Federation of American Societies for Experimental Biology 34(3):506.

Hoel, D.G., D.W. Gaylor, R.L. Kirschstein, U. Saffiotti, and M.A. Schneiderman (1975) Estimated risks of irreversible delayed toxicity. Journal of Toxicology and Environmental Health 1:133-151.

Holden, C. (1976) Mirex: Persistent pesticide on its way out. Science 194(4262):301-303.

Holdrinet, M., R. Frank, R.L. Thomas, and L.J. Hetling (In press) Mirex in Sediments of Lake Ontario. Journal of Great Lakes Research.

Huber, J.J. (1965) Some physiological effects of the insecticide Kepone in the laboratory mouse. Toxicology and Applied Pharmacology 7:516-524.

Hunter, C.G., J. Robinson, and M. Roberts (1969) Pharmacodynamics of dieldrin (HEOD). Part II. Ingestion by human subjects for 18-24 months, and post-exposure for 8 months. Archives of Industrial Health 18:12-21.

Hurlbert, S.H., M.S. Mulla, and H.R. Wilson (1972) Effects of an organophosphorus insecticide on the phytoplankton, zooplankton, and insect populations of freshwater ponds. Ecology 42:269.

Industrial Bio-Test Laboratories, Inc. (1975) Reports to Hooker Chemicals and Plastics Corp., April 18, June 13, and September 29.

Industrial Bio-Test Laboratories, Inc. (1977) Mutagenicity of PCL-HEX Incorporated in the Test Medium Tested Against Five Strains of Salmonella typhimurium and as a Volatilate Against Tester Strain TA-100. Report to Velsicol Chemical Corporation, August 10, 1977. IBT No. 8536-10838. Northbrook, Ill.: Industrial Bio-Test Laboratories, Inc.

Ingle, L. (1953) Toxicity of chlordane vapors. Science 118:213-214.

Innes, J.R.M., B.M. Ulland, M.G. Valerio, L. Petrucelli, L. Fishbein, E.R. Hart, A.J. Pallota, R.R. Bates, H.L. Falk, J.J. Gart, G.M. Klein, I.L. Mitchell, and J. Peters (1969) Bioassay of pesticides and industrial chemicals for tumorigenicity in mice: A preliminary note. Journal of the National Cancer Institute 42(6):1101-1114.

Ivie, G.W., J.R. Gibson, H.E. Bryant, J.J. Begin, J.R. Barnett, and H.W. Dorough (1974) Accumulation, distribution, and excretion of mirex-14C in animals exposed for long periods to the insecticide in the diet. Journal of Agricultural and Food Chemistry 22:646-653.

Jaeger, R.J. (1976) Kepone chronology. Science 193:94-96.

Jelinek, C. and P.E. Corneliussen (1976) Levels of PCBs in the U.S. food supply. Pages 147-154, National Conference on Polychlorinated Biphenyls, November 19-21, 1975, Chicago, Illinois. EPA-560/6-75-004. Washington, D.C.: U.S. Environmental Protection Agency.

Kellogg, R.L. and R.V. Bulkley (1976) Seasonal
concentrations of dieldrin in water, channel catfish,
and catfish-food organisms, Iowa, 1971-73. Pesticides
Montoring Journal 9:186-194.

Khera, K.S., D.C. Villeneuve, G. Terry, L. Panopie, L. Nash,
and G. Trivett (1976) Mirex: A teratogenicity, dominant
lethal, and tissue distribution study in rats. Food and
Cosmetic Toxicology 14:25-29.

Kobylinski, G.J. and R.J. Livingston (1975) Translocation of
mirex from sediments and its accumulation by the
hogchocker, Trinectes maculatus. Bulletin of
Environmental Contamination and Toxicology 14:692-698.

Koenig, C.C. (1977) The effects of DDT and mirex alone and
in combination on the reproduction of a salt marsh
cyprinodont fish, Adinia xenica. Pages 357-376,
Physiological Responses of Marine Biota to Pollutants,
edited by F.J. Vernberg, A. Calabrese, F.P. Thurberg,
and W.B. Vernberg. New York: Academic Press, Inc.

Kutz, F.W. and S.C. Strassman (1976) Residues of
polychlorinated biphenyls in the general population of
the United States. Pages 139-143, National Conference on
Polychlorinated Biphenyls, November 19-21, 1975,
Chicago, Illinois. EPA-560/6-75-004. Washington, D.C.:
U.S. Environmental Protection Agency.

Kutz, F.W., A.R. Yobs, and S.C. Strassman (1976)
Organochlorine pesticides in human adipose tissue.
Bulletin of the Society of Pharmacology and
Environmental Pathology 4(1):17-19.

Lawless, E.W., R. von Rumker, and T.L. Ferguson (1972) The
Pollution Potential in Pesticide Manufacturing. Prepared
for the U.S. Environmental Protection Agency by Midwest
Research Institute. NTIS PB-213, 782/3. Springfield,
Va.: National Technical Information Service.

Livingston, R.J. (1977) Review of current literature
concerning the acute and chronic effects of pesticides
on aquatic organisms. Critical Reviews in Environmental
Control 7:325-351. Cleveland, Ohio: CRC Press, Inc.

Livingston, R.J., N.P. Thompson, and D.A. Meeter (In press)
Long-term variation of organochlorine residues and
assemblages of epibenthic organisms in a shallow North
Florida (U.S.A.) estuary. Marine Biology.

Lowe, J.I., R.R. Parrish, A.J. Wilson, Jr., P.D. Wilson, and
T.W. Duke (1971) Effects of mirex on selected estuarine
organisms. Pages 171-186, Transactions of the Thirty-
Sixth North American Wildlife and Natural Resources

Conference, edited by J.B. Trefetheno. Washington, D.C.: Wildlife Management Institute.

Lu, P.Y., R.L. Metcalf, A.S. Hiriwe, and J.W. Williams (1975) Evaluation of environmental distribution and fate of hexachlorocyclopentadiene, chlordene, heptachlor, and heptachlor epoxide in a laboratory model ecosystem. Journal of Agricultural and Food Chemistry 23(5):967-973.

Mantel, N. and M.A. Schneiderman (1975) Estimating "safe" levels, a hazardous undertaking. Cancer Research 35:1379-1386.

Markin, G.P., J.H. Ford, J.C. Hawthorne, J.H. Spence, J. Davis, H.L. Collins, and C.D. Loftis (1972) The Insecticide Mirex and Techniques for its Monitoring. USDA Animal and Plant Health Inspection Service, APHIS 81-3. Hyattsville, Md.: U.S. Department of Agriculture, Animal and Plant Health Inspection Service.

Matsumura, F. (1975) Toxicology of Insecticides. New York: Plenum Press.

Mirex Advisory Committee (1972) Report of the Mirex Advisory Committee to William D. Ruckelshaus, Administrator, U.S. Environmental Protection Agency, February 4, 1972. Washington, D.C.: U.S. Environmental Protection Agency.

Moriarty, F. (1975) Exposure and residues. Pages 29-72, Organochlorine Insecticides: Persistent Organic Pollutants, edited by F. Moriarty. New York: Academic Press.

Morris, R.L. and L.G. Johnson (1971) Dieldrin levels in fish from Iowa streams. Pesticides Monitoring Journal 5:12-16.

Mosser, J.L., N.S. Fisher, and C.F. Wurster (1972) Polychlorinated biphenyls and DDT alter species composition in mixed cultures of algae. Science 176:533.

Mosser, J.L., T.C. Teng, W.G. Walter, and C.F. Wurster (1974) Interactions of PCBs, DDT, and DDE in a marine diatom. Bulletin of Environmental Contamination and Toxicology 12:665.

Naishtein, S. Ya. and E.V. Lisovskaya (1965) Maximum permissible concentration of hexachlorocyclopentadiene in water bodies. Gigiena i Sanitaria 30(2):17-21.

National Cancer Institute (1976) Report on Carcinogenesis Bioassay of Technical Grade Chlordecone (Kepone).

Carcinogenesis Program, Division of Cancer Cause and
Prevention. Bethesda, Md.: National Cancer Institute.

National Research Council (1975) Pest Control: An Assessment
of Present and Alternative Technologies. Volume I.
Contemporary Pest Control Practices and Prospects. Study
on Problems of Pest Control, Environmental Studies
Board, Commission on Natural Resources. Washington,
D.C.: National Academy of Sciences.

Nisbet, I.C.T. (1977) Human Exposure to Chlordane,
Heptachlor, and Their Metabolites. Report to the U.S.
Environmental Protection Agency. Washington, D.C.: U.S.
Environmental Protection Agency.

O'Connor, D.J. and K.J. Farley (1977) Preliminary Analysis
of Kepone Distribution in the James River. Environmental
Engineering and Science Program, Manhattan College,
Bronx, New York. (Unpublished report)

Peraino, C., R.J.M. Fry, E. Staffeldt, and J.P. Christopher
(1975) Comparative enhancing effects of phenobarbital,
amobarbital, diphenylhydantoin, and
dichlorodiphenyltrichloroethane on 2-
acetylaminofluorene-induced hepatic tumorigenesis in the
rat. Cancer Research 35:2884-2890.

Rajanna, B. and A.A. de La Cruz (1975) Mirex incorporation
in the environment: Phytotoxicity on germination,
emergence, and early growth of crop seedlings. Bulletin
of Environmental Contamination and Toxicology 14:77-82.

Reagan, T.E., G. Coburn, and S.D. Hensley (1972) Effects of
mirex on the arthropod fauna of a Louisiana sugarcane
field. Environmental Entomology 1:588-591.

Schimmel, S.C. and A.J. Wilson (1977) Acute toxicity of
Kepone to four estuarine animals. Chesapeake Science
18:224-227.

Smith, W.C. (1976) Kepone Discharges from Allied Chemical
Company, Hopewell, Virginia. Denver: U.S. Environmental
Protection Agency, National Field Investigation Center.
(Internal EPA memorandum)

Smith, J.C. and F.S. Arant (1967) Residues of Kepone in milk
from cows receiving treated feed. Journal of Economic
Entomology 60:925-927.

Smrek, A.L., S.A. Adams, J.A. Liddle, and R.D. Kimbrough (In
press) Pharmacokinetics of mirex in goats. 1. Effect on
reproduction and lactation. Journal of Agricultural and
Food Chemistry.

Starr, H.G., Jr., F.D. Aldrich, W.D. McDougall III, and L.M. Mounce (1974) Contribution of household dust to the human exposure to pesticides. Pesticides Monitoring Journal 8:209-212.

Sterrett, F.S. and C.A. Boss (1977) Careless Kepone. Environment 19(2):30-37.

Suta, B.E. (1977) Human Population Exposures to Mirex and Kepone. Prepared for the U.S. Environmental Protection Agency, Office of Research and Development, by the Stanford Research Institute. Contract 68-01-4314. SRI Project 5794, CRESS NO. 26. Menlo Park, Calif.: Stanford Research Institute.

Task Force on Mirex (1977) Mirex in Canada. Prepared for the Environmental Contaminants Committee of Fisheries and Environment Canada, and Health and Welfare Canada, by Task Force on Mirex. Technical Report 77-1. Ottawa, Ontario: Environmental Protection Service.

Tessari, J.D. and D.L. Spencer (1971) Air sampling for pesticides in the human environment. Journal of the Association of Official Analytical Chemists 54:1376-1382.

Treon, J.F., F.P. Cleveland, and J. Cappel (1955) The toxicity of hexachlorocyclopentadiene. Archives of Industrial Health 11:459-472.

Ulland, B.M., N.P. Page, R.A. Squire, E.K. Weisburger, and R.L. Cypher (1977) A carcinogenicity assay of mirex in Charles River CD rats. Journal of the National Cancer Institute 58:133-140.

Ungnade, H.E. and E.T. McBee (1958) The chemistry of perchlorocyclopentadienes and cyclopentadienes. Chemical Reviews 58:240-254.

United Fruit Company (1969) Petition for the Establishment of Tolerances for the Pesticide Chemical Decachlorooctahydro-1, 3, 4-metheno-2H-cyclobuta [cd] pentalen-2-one on Raw Bananas. Submitted to the U.S. Environmental Protection Agency as part of Pesticide Petition No. OE0919. Washington, D.C.: U.S. Environmental Protection Agency. (Unpublished data)

U.S. Department of Agriculture (1977) Comments of the Secretary of Agriculture in Response to the Notice of Intent to Cancel Pesticide Products Containing Chlordecone, Trade Name Kepone. January 11, 1977. Washington, D.C.: U.S. Department of Agriculture.

U.S. Environmental Protection Agency (1975) DDT: A Review of Scientific and Economic Aspects of the Decision to Ban its Use as a Pesticide. EPA-540/1-75-022. Washington, D.C.: U.S. Environmental Protection Agency.

U.S. Environmental Protection Agency (1976a) Analysis of Kepone. Prepared by the Carcinogen Assessment Group. Washington, D.C.: U.S. Environmental Protection Agency.

U.S. Environmental Protection Agency (1976b) EPA finds Kepone in human milk in the South. Environmental News, February 27, 1976. Washington, D.C.: U.S. Environmental Protection Agency.

U.S. Environmental Protection Agency (1976c) Human Monitoring Program. Mirex Special Report. Pesticide Monitoring Quarterly Report 5:2-11.

U.S. Environmental Protection Agency (1976d) Review of the Chesapeake Bay Program Seminar on Kepone held at Virginia Institute of Marine Sciences, October 12-13, 1976. Washington, D.C.: U.S. Environmental Protection Agency. (Information memorandum)

U.S. Environmental Protection Agency (1976e) Notice of intent to cancel registrations of pesticide products containing chlordecone (Kepone). 41 FR 2464.

U.S. Environmental Protection Agency (1977a) Establishment and Reviews of Original Kepone Action Levels. Office of Pesticide Programs, Registration Divison, January 10, 1977. Washington, D.C.: U.S. Environmental Protection Agency.

U.S. Environmental Protection Agency (1977b) National Study to Determine Levels of Chlorinated Hydrocarbon Insecticides in Human Milk (1975-76). Contract No. 68-01-3190. Washington, D.C.: U.S. Environmental Protection Agency.

U.S. Environmental Protection Agency (1978) Review of the Environmental Effects of Mirex and Kepone. Prepared for the U.S. Environmental Protection Agency, Office of Research and Development, by Battelle Columbus Laboratories (M.A. Bell, R.A. Ewing, and G.A. Lutz). EPA 600/1-78-013. Washington, D.C.: U.S. Environmental Protection Agency.

U.S. Food and Drug Administration (1971) Cancer Testing in the Safety Evaluation of Food Additives and Pesticides. Report of the Advisory Committee on Protocols for Safety Evaluation, Panel on Carcinogenesis. Toxicology and Applied Pharmacology 20:419-438.

Velsicol Chemical Corporation (1976) Product Bulletin Number 50101-2, April 8, 1976. Chicago: Velsicol Chemical Corporation.

Walsh, G.E., K. Ainsworth, and A.J. Wilson (1977) Toxicity and uptake of Kepone in marine unicellular algae. Chesapeake Science 18:222-223.

Ware, G.W. and E.E. Good (1967) Effects of insecticides on reproduction in the laboratory mouse. II. Mirex, Telodrin, and DDT. Toxicology and Applied Pharmacology 10:54-61.

Whetstone, R.R. (1964) Chlorinated derivatives of cyclopentadiene. Pages 240-252, Kirk-Othmer Encyclopedia of Chemical Technology, Volume 5, Second Edition. New York: Interscience.

Wiener, M., K.A. Pittman, and V. Stein (1976) Mirex kinetics in the rhesus monkey. I. Disposition and excretion. Drug Metabolism and Disposition 4:281-287.

Yobs, A.R. (1971) The national human monitoring program for pesticides. Pesticides Monitoring Journal 5:44-46.

APPENDIX

ORGANIZATIONS AND INDIVIDUALS
CONTACTED FOR PUBLIC INPUT

Government and International Organizations

*Agriculture Canada
Alabama Department of Agriculture and Industries
Arkansas Department of Agriculture
Arkansas Plant Board
*Canada Department of Fisheries and the Environment
Consumer Product Safety Commission
Council on Environmental Quality
Florida Department of Agriculture and Consumer Services
Food and Agricultural Organization, United Nations
Fort Bend County (Texas) Fire Ant Committee
Georgia Department of Agriculture
Great Lakes Commission
Hawaii Department of Agriculture
Louisiana Department of Agriculture
*Michigan Department of Natural Resources
Mississippi Department of Agriculture and Commerce
*National Cancer Institute
*National Institute of Environmental Health Sciences
National Institute of Occupational Safety and Health
National Institutes of Health
*New York Department of Health
North Carolina Department of Agriculture
North Carolina Department of Natural and Economic
      Resources
North Carolina Pesticides Board
Occupational Safety and Health Administration
Organization for Economic Cooperation and Development
*Organization of American States
Pan American Health Organization
Patuxent Wildlife Research Center
South Carolina Department of Agriculture
Texas Department of Agriculture
*U.S. Department of Agriculture
U.S. Department of Commerce
*U.S. Department of Health, Education, and Welfare
U.S. Department of Interior
*U.S. Environmental Protection Agency
*U.S. Fish and Wildlife Service

U.S. Maritime Administration
World Health Organization

Environmental Public Interest Organizations

Center for Science in the Public Interest
Commission for the Advancement of Public Interest
    Organizations
Ecology Center of Louisiana, Inc.
Environmental Defense Fund
Environmental Law Institute
Environmental Policy Center
Friends of the Earth
Izaak Walton League of America
National Audubon Society
National Wildlife Federation
Natural Resources Defense Council
Public Citizens Health Research Group (Nader's)
Science for the People
Scientists Institute for Public Information
Sierra Club
The Wilderness Society

Professional/Scientific/Trade Organizations

American Association for the Advancement of Science
American Chemical Society
American Institute for Biological Sciences
American Law Institute
American Medical Association
American National Cattlemen's Association
American Public Health Association
American Society for Testing and Materials
American Society of Limnology and Oceanography
Association of American Pesticide Control Officials,
    Inc.
Chemical Industry of Toxicology
Chemical Specialties Manufacturers Association
*Council for Agricultural Science and Technology
Ecological Society of America
Entomological Society of America
Federation of Amercian Scientists
*Hawaii Pineapple Growers Association
Manufacturers Chemists Association
National Agricultural Chemicals Association
National Association of Manufacturers
National Association of State Departments of Agriculture
National Environmental Health Association
National Pest Control Association
National Research Council, Canada
National Science Foundation
Resources for the Future

70

Society of Occupational and Environmental Health
Society of Toxicology
Soil Science Society of America
Sport Fishing Institute
Synthetic Organic Chemicals Manufacturers Association
Texas Farm Bureau
The Ecosystem Center, Woods Hole, Mass.
The Institute of Ecology
Water Pollution Control Federation

Consultant/Industry Organizations

*Allied Chemical Corporation
Arthur D. Little, Inc.
*Battelle Columbus Laboratories
*C.F. Spiess and Sohn, Federal Republic of Germany
Dow Chemical Company
*Equitable Environmental Health, Inc.
*Hooker Chemicals and Plastics Corporation
Litton Bionetics, Inc.
McGuire, Woods, and Battle
Midland-Ross Corporation
Midwest Research Institute
Shell Chemical Company, Holland
*Stanford Research Institute
United Fruit Company
*Velsicol Chemical Corporation
Zoecon Corporation

Individuals

Earl G. Alley
Claus Agthe
Karl Baetcke
William A. Banks
John W. Beardsley, Jr.
*Jack Blanchard
Norman E. Blomberg
Cazlyn G. Bookhout
Joe A. Bowden
Samuel M. Branco
Ralph E. Brown
J.W. Butcher
Gerald A. Carlson
Neil Chernoff
*K. Christiansen
Roy P. Clark
Nelson R. Cooley
*D.A. Crossley, Jr.
*Delamare Deboutteville
Justin Dillier
Marden Dixon

*Alexander M. Dollar
*K.H. Domsch
Gus R. Douglas
*Thomas W. Duke
Donald Eichler
Daniel R. Embody
Joseph H. Ford
*Richard Frank
John H. Gart
B. Michael Glancey
T.R.G. Gray
Adrian Gross
Alva Harris
*S.D.Hensley
*Ernest Hodgson
Terry A. Holister
William T. Holloway
John W. Hurdis
*G. Iagnov
David Ivie
Jerome A. Jackson
*Rudolph J. Jaeger
Frank K. James
Jerry M. Johnson
Halwin L. Jones
Kingsly Kay
James Y. Kim
*Renate D. Kimbrough
Edwin V. Komarek
Paul Kotin
Bob Krieger
F.W. Kutz
*John L. Laseter
*E. Paul Lichtenstein
Jeffrey L. Lincer
*Charles G. Lincoln
*J.L. Lockwood
Clifford S. Lofgren
J.I. Lowe
Nathan Mantel
George P. Markin
William H. Matheny
*H.M. Mehendale
*G Minderman
*M.M. Mortland
*Donald Mount
Sheldon Murphy
*Paul O. Nees
Leo D. Newsome
David Nobriga
Watson T. Okubo
Frederick W. Plapp, Jr.
Bart J. Puma
Melvin D. Reuber

72

Kanjyo Sakimura
Hope Sandifer
Louis H. Senn
*William F. Spencer
William W. Stearns
Jean Stellman
Martin E. Tagatz
Rayford O. Thompson
*James Tiedje
*Ernest L. Timlin, Jr.
Walter R. Tschinkel
Marion Ueltschey
J. van der Drift
S. Bradley Vinson
*Joe L. White
Robert A. Williamson
Alfred J. Wilson, Jr.
George Woehr

* Parties that responded to Panel's request for public
input; copies of written replies are on file for inspection
at the Environmental Studies Board, Commission on Natural
Resources, National Academy of Sciences.